WEATHER

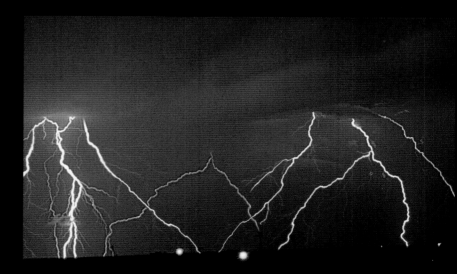

WEATHER

Julie Lloyd

This is a Parragon Book
First published in 2007

Parragon
Queen Street House
4 Queen Street
Bath, BA1 IHE

For photograph copyrights see page 320
Text © Parragon Books Ltd 2007

Produced by Atlantic Publishing

A catalogue record for this book is available
from the British Library.

ISBN: 978-1-4054-8796-2

Printed in China

CONTENTS

Introduction

For most of us the weather is only an issue when it comes to deciding each morning what to wear, or perhaps when considering where to take a holiday. Is it cold or hot? Will it rain or be windy? At other times it becomes a consideration only when there is something unusual taking place, such as a period of extreme heat or a storm. It is perhaps the most common topic of conversation in many places. Increasingly, we have become interested in the weather as we read about and experience more unusual effects.

Had Captain Edward Smith understood how El Niño turns global weather patterns on their head, Titanic may have survived its maiden voyage. In 1912 the planet was in the grip of this cyclical phenomenon, which can produce parched rainforests and verdant deserts. It was also responsible for the unusually powerful northerlies which drew the iceberg much further south than usual, with catastrophic consequences for the 'unsinkable' liner.

Meteorology is a young science. It is only in the last two centuries that we have learned how the sun's energy acts on the Earth's atmospheric envelope to produce the planet's diverse weather episodes and climatic zones. We no longer believe, as primitive peoples did, that thunderstorms, rainbows and other atmospheric phenomena are omens sent by sky deities. We try to control the weather, to harness its extraordinary forces, yet we are still in thrall to Nature's climatological handiwork.

This book demystifies atmospheric phenomena, from haloes to hurricanes, sundogs to supercells. It also addresses the issue of climate change, which has brought geography to the forefront of the political debate. To what extent is mankind responsible for the fact that 19 of the hottest 20 years on record have occurred since 1980? Even if carbon dioxide emissions were to be drastically curtailed, has the tipping point already been reached?

Global temperatures have fluctuated considerably over the Earth's 4600 million-year history, and there is evidence to suggest that five climate-related mass extinctions of animal and plant life have occurred. It is a situation of few certainties, though one thing is clear: if Captain Smith were undertaking his voyage today, his vessel would complete its Atlantic crossing unhindered: the waters where Titanic foundered are now too warm to sustain icebergs.

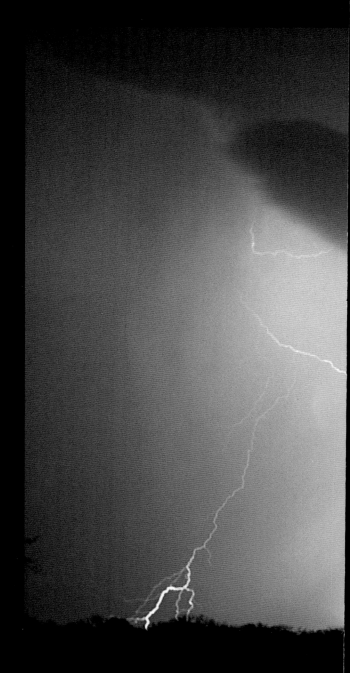

CLIMATE ZONES

WEATHER AND CLIMATES

The Earth's atmosphere is constantly in motion and it is this which creates our weather. Weather is usually defined as the current state of the atmosphere in a place and time. The weather is, arguably, generally looked at as being of a relatively small scale both in terms of area and of time. Weather forecasts are prepared frequently, on at least a daily basis for a given location.

In contrast, climate is seen as the longer term general state of the atmosphere; if you like it is the typical weather for a location. It takes into consideration long-term averages, usually over a minimum of 30 years. This is designed to eliminate the effects of 'freak' events and so give a clearer picture of what to expect.

Meteorology, the scientific study of the weather, has developed significantly during the last 50 years but we have reliable records for at least two centuries. There is a significant body of supplementary information, however, from a variety of historical sources which is now being explored to enable us to have a much clearer picture of the past. This might help us to better understand not only the past but also predict the future.

Climate Zones

The average weather conditions of a place are called its climate. The most important conditions are the average monthly temperatures and average monthly rainfall. The world is divided into regions according to their climate, although the boundaries are not sharply defined, as one type of climate gradually merges into the next. In addition every year is different; for example, the occasional long hot summers over north-west Europe are called Mediterranean type summers because they more nearly reflect that climate.

The first person to define the climatic zones was the Greek philosopher Aristotle (384–322 BC). He divided the world into three zones (torrid, temperate and frigid) based on the height of the sun above the horizon. Modern climatic maps show many more regions using a variety of criteria including latitude, height above sea level, distance from the ocean, temperature and precipitation. In broad terms, climate zones range in latitude from tropical (hot; no real seasons based on temperature) to mid-latitude (moderate; conventional winter-spring-summer-autumn seasons) to polar (cold; strong seasonality; long winters).

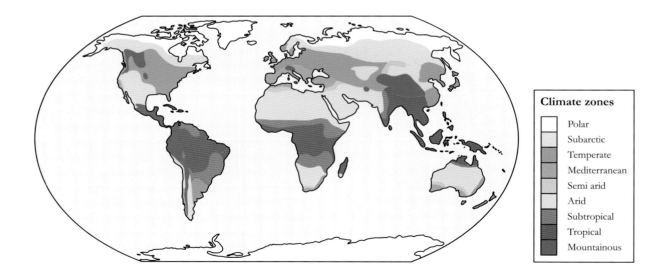

Climate zones

Polar
Subarctic
Temperate
Mediterranean
Semi arid
Arid
Subtropical
Tropical
Mountainous

Above: Tropical beach in the Seychelles. The Republic of Seychelles represents an archipelago of more than 100 tropical islands that extends between 4 and 10° south of the Equator in the Indian Ocean. The climate is a tropical haven where the temperature seldom drops below 24°C/75°F or rises above 32°C/90°F.

Right: Ice floes in Antarctica. Climates vary enormously round the world and affect the character of an area, the plants and animals that live there and the people and homes they live in. Near the Equator the climate is hot and mostly wet. Deserts are dry while the poles have ice and snow all year round.

Above: The rainforest is divided into four distinct parts. The emergent layer contains the small number of trees which grow above the general canopy. The canopy contains the majority of the larger trees; it is estimated that it is home to 40 per cent of all plant species. The understory layer is the space between the canopy and the forest floor and is home to numerous birds, snakes and lizards. The forest floor is relatively clear of vegetation and contains mainly decaying plant and animal matter.

Above: Rainforests are home to two-thirds of all the living animal and plant species on the planet. The woolly spider monkey is native to the Atlantic rainforests of Brazil, but is now endangered as a result of hunting and habitat loss.

Opposite top: In coastal California, summers are generally hot, and winters are mild and humid. Summer temperatures can exceed 32°C/90°F, and smog can become a problem. In winter, temperatures dip to an average of 12°C/55°F and rain is a possibility.

Tropical

The tropics lie between the Tropic of Cancer and the Tropic of Capricorn (latitude 23.5° north and south). It is only in this zone that the sun is directly overhead with the result that temperatures are always high (except in highland areas) and the seasons are marked only by changes in wind and rainfall.

These conditions are ideal for vegetation and consequently the greatest rainforests in the world lie within 10° of the Equator. The most well known are the Amazon rainforest of Brazil, the Congo basin of Central Africa and the jungles of Malaysia, Indonesia, Burma and Vietnam.

In moist tropical rainforests it rains nearly every day and is hot and humid throughout the year. In the interior it is damp and dark and the trees grow up to 60 metres/196 feet tall as they compete to reach the light and form the well-known rainforest canopy. Under this layer are smaller trees, masses of creepers and rope-like lianas hanging from the trunks and branches trying to reach the light.

The layered structure of the rainforest gives different amounts of light from top to bottom where thousands of plants thrive and support a huge variety of animal, bird and insect life.

Subtropical

The subtropics refers to the zones of the Earth immediately north and south of the Tropic of Cancer and the Tropic of Capricorn, roughly between 23.5° and 35° north and south. These areas typically have very warm to hot summers, but non-tropical winters. A subtropical climate implies that the air temperature usually does not fall below freezing (0°C/32°F) and includes the regions of coastal California, southern Florida, northern India, coastal Australia, Texas and coastal South Africa. Subtropical regions don't usually have distinctly wet or dry seasons, and have a fairly even distribution of rain throughout the year.

Left: The acacia tree (sometimes called the camelthorn) grows in the deserts of Africa. It has developed ways to survive the hot and dry climate in which it lives. Its leaves have evolved into many tiny leaflets which can turn to either absorb maximum sunlight, or to avoid it and reduce transpiration.

Below: Namib Desert Dunes, Sossusvlei, Namibia. There is very little precipitation (less than 10 millimetres/0.4 inches per year) in the Namib Desert apart from rare thunderstorms. The only other moisture comes from advection fog or mist. Its aridity is caused by the descent of dry air cooled by the cold Benguela current along the coast of Namibia.

Arid

Hot deserts and semi-desert regions occupy one-fifth of the world's land surface. Deserts are dry, barren places that receive an average annual precipitation of less than 250 millimetres/10 inches. They usually have an extreme temperature range and can have high daytime temperatures of over 50°C/122°F due to the intense heat of the sun and the lack of clouds. Even though the desert is very hot in the day, it is extremely cold at night because the air is very dry and therefore holds little heat, so as soon as the sun sets, the desert cools quickly. Also, cloudless skies increase the release of heat at night.

Deserts are made in different ways, but are always formed because there is not enough water.

Desert landscapes have certain distinctive features and are often composed of sand and rocky surfaces. There are three main types: sandy deserts or ergs, rocky deserts or hammada and stony deserts or reg. Deserts also sometimes contain valuable mineral deposits formed in the arid environment or exposed by erosion. Because deserts are dry, they are ideal places for human artifacts and fossils to be preserved.

Deserts have a reputation for supporting very little life, but in reality they are often home to a high diversity of animal life; many of these remain hidden during daylight hours to control body temperature or to limit moisture needs. Rare showery spells can initiate the life cycle of dormant insects such as the avaricious locust, the eggs of which sometimes lie for years in the dry sand.

Semi-arid (temperate grasslands)

There are vast areas of grassland in the interiors of the continents in the temperate zone. They develop where there is too little precipitation for trees to grow and the climate can be summarized by hot summers and cold winters, averaging 250–500 millimetres/10–20 inches of rain or equivalent in snowfall a year. The soil is considered too moist to be a desert, but too dry to support normal forest life.

Various names are given to these grasslands, including prairie, steppe and pampas. These areas may be semi-desert, or covered with grass or shrubs, or both depending on the season and latitude.

The world's largest zone of steppes, often referred to as 'the Great Steppe', is found in central Russia and neighbouring countries in Central Asia and stretches for 3000 kilometres/1860 miles. The pampas are in Argentina, Uruguay and Brazil, stretching from the foothills of the Andes to the Atlantic. The veld in the eastern corner of South Africa is close to the sea and does not suffer the same extremes of temperature. Australia's temperate grasslands occupy most of the basin of the Murray-Darling River. A significant steppe, noteworthy for not grading into desert, is the Sertão of north-eastern Brazil.

The North American prairies are one of the most famous examples of temperate grasslands. These reach from Canada's prairie provinces almost to the Gulf of Mexico, and east to south of the Great Lakes. Herds of pronghorn antelopes and bison used to feed on these prairies before they were hunted close to extinction by 19th-century settlers. Today most of the land is planted with wheat, oats, barley and maize.

Many kinds of natural grasslands are now becoming rare. About two-thirds of the Canadian prairies and half the United States prairies have been converted into huge cereal farms and there is little natural vegetation left. Much of the South African veld has also been destroyed.

Above: The American bison, or buffalo, is the largest terrestrial mammal in North America. Bison are nomadic grazers and travel in herds on the Great Plains – the broad expanse of prairie and steppe which lies east of the Rocky Mountains in the United States and Canada.

Mediterranean (warm temperate)

Despite its name this climate is not peculiar to countries fringing the Mediterranean Sea. There are Mediterranean climates between 30° and 40° north and south of the Equator, for example parts of the USA including California, much of South Africa and southern Australia, northern parts of Chile and Argentina and much of the North Island, New Zealand.

During the summer, warm dry winds blow from the Equator, bringing dry conditions. In the winter the winds swing towards the Equator and westerly winds bring rain. Summers are hot, and winters are usually warm enough for plants to grow. Frost and snow are rare except over mountains.

In the area around the Mediterranean, the sea has a big effect on the climate. In the summer it is cooler than the land, so the air sinks down over the sea and the surrounding area and there is very little rain. Few clouds mean high temperatures. In the winter the sea is warmer than the land, so the surrounding land enjoys mild winters while the warm, moist air from the sea brings rain. In other parts of the world with this climate, cold ocean currents affect the cold climate.

The Mediterranean climate is ideal for growing many fruits, especially grapes, melons and citrus and many of these areas are important wine producers. Olives have been grown in these regions for centuries and provide oil in areas where milk and butter are hard to produce. Vegetables also thrive, and often more than one crop a year is possible, especially where irrigation is available to supplement the meagre summer rainfall. Tourism is also a major industry in these climate regions.

Above left: Tourism has become a major industry for many areas bordering the Mediterranean Sea. The hot, dry summers have made these places popular holiday destinations for northern Europeans keen to take advantage of the Mediterranean climate.

Left: The grape vine, with its long roots and tough bark is well suited to the Mediterranean climate, and many vineyards are found in these areas.

Temperate (cool temperate)

The temperate regions of the world lie between the tropics and the polar circles – in the latitudes of the changeable westerlies. The changes in these regions between summer and winter are generally subtle, warm or cool, rather than extreme, burning hot or freezing cold. However, a temperate climate can have very unpredictable weather. One day it may be sunny, the next it may be raining, and after that it may be cloudy. These erratic weather patterns occur in summer as well as winter.

Weather systems bring periods of rain throughout the year, which penetrate well into the interior of continents such as Europe which has no mountain barrier. Temperate climates include northern Europe, British Columbia, Oregon and Washington State in North America, southern Chile, Tasmania and South Island, New Zealand.

In temperate zones latitude and the shelter provided by hills or mountains have the most marked effect on average temperature and rainfall. However, all these regions have a dormant season when the temperature is below 6°C/43°F and plant growth ceases. Conditions are generally favourable for abundant plant and animal life and approximately 80 per cent of the world's population live in temperate zones.

Above right: Tsitsikamma along South Africa's southern coastline has a temperate coastal climate with an annual rainfall of 1200 millimetres/47 inches. The wettest months are May and October, and the driest are June and July.

Right: Temperate climates support a wide variety of plant and animal life due to the mild climate, plentiful rainfall and generally fertile soil. Although temperate zones only account for about seven per cent of the world's land surface, they are by far the most popular areas in which to live and are home to around four-fifths of the world's population.

Subarctic

This climate is peculiar to the Northern Hemisphere where the largest continents of the world extend to the Arctic. Regions having a subarctic climate are characterized by long, very cold winters, and brief, warm summers. This type of climate offers some of the most extreme seasonal temperature variations found on the planet: in winter, temperatures can drop to −40°C/−40°F and in the short, hot summers the temperature may exceed 30°C/86°F. With six to seven consecutive months where the average temperature is below freezing, all moisture in the soil and subsoil freezes solidly to depths of many metres. The frost-free season is very short and the warmth of summer is insufficient to thaw more than a few surface metres, so permafrost prevails under large areas.

Vegetation in the subarctic climate consists almost entirely of conifers, which survive extreme cold and are not easily damaged by snow. Animal life has been plentiful in the past, particularly large deer, brown bears and wolves, but indiscriminate hunting in recent years has resulted in a decline in the number of these animals.

Above: Mountains give rise to distinctive climates. Temperatures are lower in elevated environments because the air is further from the Earth's surface.

Below: Only six out of the 35 species of seal live in Antarctica. These crabeater seals live at the edge of pack ice and, despite their name, eat mainly krill. Seals are streamlined, highly specialized marine creatures that have adapted to life in the cold waters of the Antarctic climate with a thick layer of blubber and fur.

Polar

The harshest climate on Earth occurs within the Arctic and Antarctic circles. These regions are cold, icy deserts where it is bitterly cold all year round. The lowest temperatures recorded are in these areas – the record being –89°C/–129°F at Vostock, a Russian research station in the Arctic, in July 1983. While high pressure is often maintained for long stretches of time, deep depressions bring long periods of snow, strong winds and blizzards at all times of the year.

The Antarctic region around the South Pole includes a huge continent almost covered with ice and snow. A deep ocean partly covered by floating ice surrounds it, even in summer. Antarctica has such a harsh climate that people have never settled there permanently. It is by no means deserted, however, as many scientists visit the region to carry out research projects and there are more than 40 research stations operating all year round.

The northern polar region is an ocean basin surrounded by land. Floating ice covers much of the Arctic Ocean all year round with ice so thick that even the strongest ships cannot reach the North Pole. In warmer parts of the Arctic lands, ice and snow disappear in summer and tundra plants grow on the stony ground. The tundra is an area where tree growth is hindered by low temperatures and short growing seasons and the dominant vegetation is grasses, mosses, and lichens.

During the summer months in the polar regions, many types of small plants and millions of insects enjoy their brief life cycle, especially in the Arctic as migratory birds fly in. Year-round natural inhabitants are mostly restricted to seals, sea lions and penguins in the south, and polar bears in the north. There is no farming, but fishing grounds yield high catches in the summer as the ice retreats.

Mountain

Mountains interrupt the flow of air around the Earth's surface and give rise to distinctive climates. In general mountain regions are characterized by cooler temperatures, higher winds and more precipitation than the surrounding area. Temperatures are lower in elevated environments because the air is further from the Earths surface, which is heated by solar radiation. The colder temperature means that precipitation often falls as snow in the mountains. Because they can be found at varying latitudes, mountain climates often differ from one another; a mountain on the Equator will experience a different weather patterns from one nearer the poles.

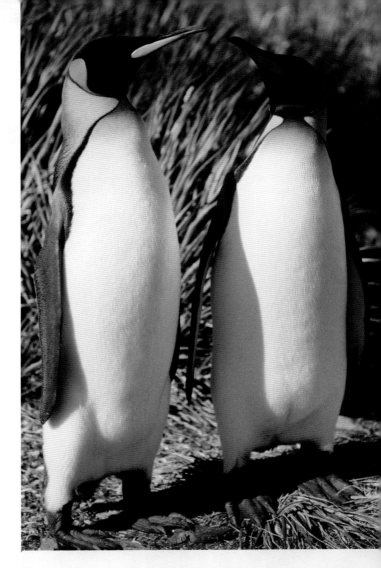

Above: Antarctica is an isolated continent that is no longer the home to any terrestrial animals. However, there is a diverse and abundant community of animals living within the Antarctic shelf. Several varieties of penguin live on the ice around the edge of the Antarctic peninsula as well as the larger icebergs that break off from the Antarctic ice. To withstand the harsh conditions of the Antarctic, their bodies are insulated by a thick layer of blubber and a dense network of waterproof plumage.

THE ATMOSPHERE: GLOBAL PATTERNS

WHAT CAUSES WEATHER?

The weather we experience at the Earth's surface is caused by a complex set of interrelated factors. Some of these are large-scale, affecting all parts of the Earth at some, or for most of the time. These include the impact of solar heating – energy from the sun and the tilt of the Earth on its axis in relation to the sun, in other words the seasons. Other large-scale, global factors include the amount of water vapour and other gases in the atmosphere which influence global temperatures. Additionally, the impact of high and low pressure systems and large scale permanent or periodic ocean currents is of fundamental importance in explaining why we have particular weather and climate.

Other factors affect only relatively small areas of the globe and are more localized in their impact. These are sometimes called meso- or, even smaller, micro-scale effects. These may include the impact of altitude – it becomes cooler and wetter with altitude; local winds such as the hot Santa Ana wind of California or the cooler Mistral of France or, whether the area is rural or urban – urban areas can develop their own micro-climates. Other local aspects of physical geography, such as valleys or coasts can have an impact on local weather: river valleys may experience fog more frequently than other areas and coastal areas generally experience cooler temperatures in summer and milder in winter than areas further inland. They may also be windier due to the impact of sea and land breezes.

In addition to the impact of the natural environment anthropogenic factors – human activity – are now of increasing concern. The huge amount of publicity given to global warming and climate change has made us more aware of the impact humans can have on the weather and climate we experience.

Above: A rainbow can be observed whenever there are water drops in the air and sunlight shining from behind the observer.

Opposite: Sunset over the flooded pan, in Atosha, Namibia. The colour of a sunset may be enhanced by atmospheric phenomena such as clouds, smoke and smog produced by natural processes or human activity, and by ash from volcanic eruptions.

Right: Lightning is a powerful, natural electrostatic discharge produced during a thunderstorm. The abrupt electrical discharge is accompanied by the emission of visible light and other forms of electromagnetic radiation.

THE SUN

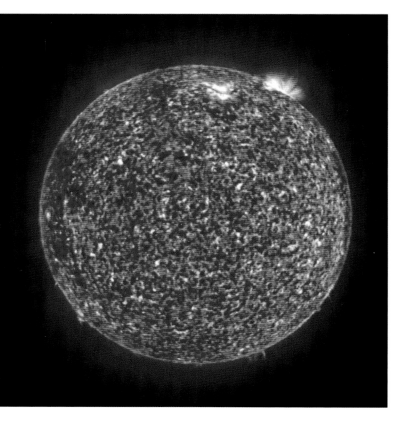

Solar Energy

The sun provides the energy which drives the Earth's weather systems. The energy comes from the thermonuclear reactions taking place in its core. These can give rise to temperatures as high as 6000°C /11,000°F on the surface. The reactions produce energy which naturally radiates in all directions; only a small but vital part is intercepted by the Earth as it orbits around the sun. This takes 365.25 days – the fraction is, of course, rounded up every four years in leap years. The Earth's orbit is elliptical (oval shaped), which means that the Earth's distance from the sun may vary by as much as 4,800,000 kilometres/3 million miles.

About 46 per cent of the sun's energy is in the form of visible light. A similar amount is in the form of infrared radiation – heat. The remainder is ultraviolet radiation, the cause of sunburn in humans.

Additionally there are solar eruptions called prominences and flares which occur periodically and give off considerably more energy and are believed to have significant temporary effects on our weather.

Above: The sun is a huge ball of hydrogen gas, ionized into a plasma by the immense temperatures that are generated by nuclear fusion at its core. This fusion is ignited by the immense pressure generated by the sun's mass. An active sun is more likely to eject plasma from its surface and the level of activity varies, peaking roughly every 11 years. The sun's light and heat directly or indirectly provides almost all of the energy for life on Earth although it is some 150 million kilometres/93 million miles distant from Earth.

Right: Cutaway illustration showing the layers of the sun. At the sun's core (blue), hydrogen atoms undergo nuclear fusion, producing helium atoms and releasing heat and light energy. The energy travels as photons through the inner region (green-yellow) and then rises by convection through the outer region (orange). The sun's surface (photosphere) has a temperature of about 5700 Kelvin (K), 5500°C/9800°F, compared with 15 million K (15 million °C/27 million °F at the core. Turbulent magnetic fields cause surface phenomena like sunspots (cooler, darker areas, brown, lower left) and solar prominences (eruptions of charged particles, upper right).

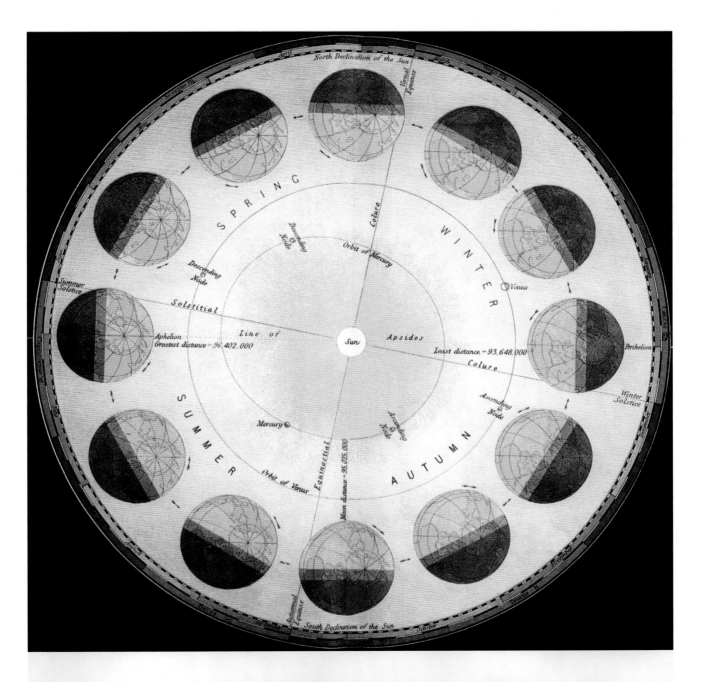

Above: Historical artwork showing the Earth revolving around the sun. The Earth takes around 365 days to complete a full orbit of the sun. The orbit is not perfectly circular, but elliptical. The Earth is tilted on its axis as it rotates, and this gives rise to the changing seasons. When the North Pole points away from the sun it is winter in the Northern Hemisphere, and when it points towards the sun it is summer in the Northern Hemisphere. This picture is from *The Popular Encyclopedia; or, Conversations Lexicon*, which was published around 1879.

Seasons

As the Earth makes its orbit around the sun there are changes in the amount of sunlight received which affects the temperatures experienced.

The Earth spinning on its axis makes a single rotation every 24 hours, in other words, it creates a day and a night. It takes 365.25 days for Earth to complete a single orbit of the sun. As the Earth is tilted on its axis – 23.5° – different parts of the world receive different amounts of sunlight at different times of year. At the Equator there is little variance in sunlight and so in length of day and in temperatures. This means there are no significant seasons.

The Earth's tilt means that while it orbits the sun, the Northern and Southern Hemispheres are oriented further towards and further away from the sun, with the result that the sun appears higher or lower in the sky, for longer or shorter periods, and its rays are experienced more or less directly or obliquely. In the Northern Hemisphere the longest day, the summer solstice, is on 21 June. This coincides with the winter solstice or shortest day in the Southern Hemisphere. These periodic variations in the amount of sunlight received, and therefore also the temperatures experienced, are delineated as the seasons.

In the polar and temperate regions, the astronomical year is divided into four equal parts by the two solstices, when the sun is at its extreme northern and southern declinations, and the two equinoxes, when the sun traverses the celestial equator. The solstices mark the longest and shortest days of the year, whilst at the vernal and autumnal equinoxes, day and night are of equal duration.

The four seasons, which are described as spring, summer, autumn (fall), and winter, are in astronomical terms divided by the vernal and autumnal equinoxes, and by the summer and winter solstices, with each marking the end of one season and the beginning of the next. However, in meteorological terms, the seasons are taken to begin around three weeks prior to these occurrences. Importantly, it should also be noted that at any particular time of the year, each hemisphere is experiencing the opposite season,

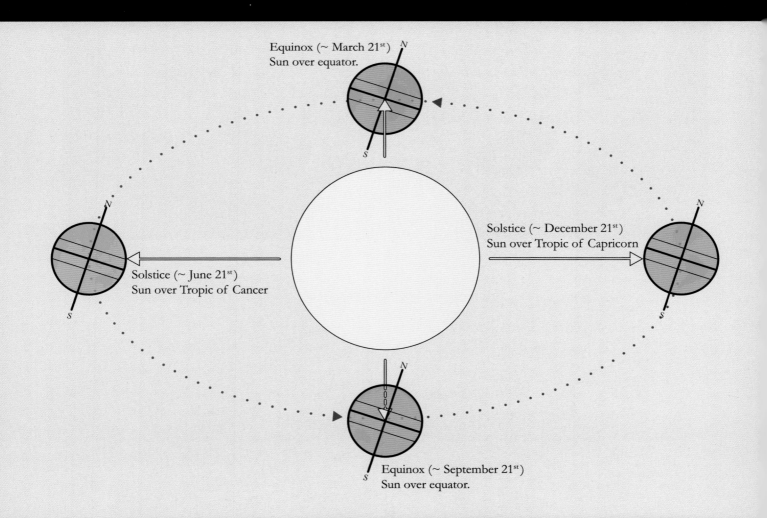

Equinox (~ March 21st) N
Sun over equator.

Solstice (~ June 21st)
Sun over Tropic of Cancer

Solstice (~ December 21st)
Sun over Tropic of Capricorn

Equinox (~ September 21st)
Sun over equator.

Above: Lavender is one of the best tempered and least demanding of plants, asking only well-drained alkaline soil and plenty of sunshine. Although a native of the Mediterranean region, it grows throughout the world where soil and climate are favourable.

Right: In many parts of the world the appearance of broadleaf trees defines the season. Naked branches in the winter become clothed in the vivid greens of spring as the young leaves, full of green chlorophyll, unfurl and grow. As the leaves mature during the summer duller hues predominate. Finally with the onset of autumn the chlorophyll is broken down to reveal other pigments hitherto masked which provide the red, bronze and yellow colours so typical of the autumn landscape.

Opposite: We have seasons because the Earth is tilted on its axis and remains tilted at the same angle as it circles the sun. For part of the year the Northern Hemisphere is tilted towards the sun, giving that part of the world the long, warmer days of summer. At the same time, the Southern Hemisphere is tilted away from the sun and has the shorter cooler days of winter. When the Earth travels to the far side of the sun, the Southern Hemisphere is tilted towards the sun and the seasons reverse.

Above: The fly agric mushroom is one of many fungi that appear during the autumn, often emerging overnight in woodland and pasture.

Below: Condensation on the web of an orb-web spinning spider. Spiders' webs are found in the autumn on plants and bushes. They are easiest to see in the morning when they are covered in dew.

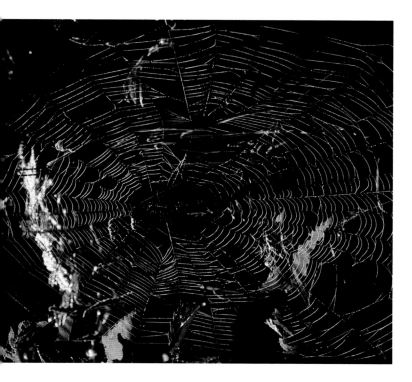

so that the longest day in the Northern Hemisphere coincides with the shortest day in the Southern Hemisphere, and vice versa.

Closer to the Equator, in tropical and subtropical regions, the degree of exposure to sunlight is relatively constant, but the movement of the tropical rain belt, which crosses the Equator from one hemisphere to the other, tends to result in two seasons: the rainy, wet, or monsoon season, and the dry season, between which the degree of precipitation may differ considerably. On the Equator itself, two rainy and dry seasons may be experienced each year.

Spring

Astronomically, spring begins in the Northern Hemisphere about 21 March, and in the Southern Hemisphere around 21 September, lasting until about 21 June and 21 December respectively.

In spring, the Earth is tilted towards the sun, resulting in increasing hours of daylight and rising temperatures. The hemisphere concerned is warmed, causing the thaw of the winter snow, a profusion of new plant growth and the blooming of many flowering plants. Many animals also give birth to their young at this time, and those that have spent the winter in hibernation awaken. Spring is regarded as a time of birth and growth, when new life literally begins to spring forth. However, the season can also bring severe weather, such as increased precipitation and high winds, resulting in storms, flooding, tornados and hurricanes.

Summer

Astronomically, summer begins in the Northern Hemisphere on about 21 June, and in the Southern Hemisphere around 21 December, lasting until about 23 September and 21 March respectively.

As in spring, the Earth is tilted towards the sun in summer, and daylight hours and temperatures continue to increase, ensuring that the longest and warmest days are experienced, and it may also be uncomfortably warm at night. Increased humidity can make it feel 'sticky' or 'close', and commonly leads to thunderstorms. Plants grow strongly on account of the high levels of sunlight, and often produce fruits, whilst many animals breed throughout the season.

Autumn

23 September is the first day of autumn in the Northern Hemisphere and in the Southern Hemisphere around 21 March, lasting until about 21 December and 21 June respectively. However, in meteorological terms, it is often taken to include September, October and November, or conversely, March, April and May, in their entirety.

Autumn, or fall, as it is commonly known in North America, marks the transition between summer and winter, when the Earth is tilted away from the sun, daylight hours shorten, and temperatures become cooler. Precipitation may also increase in many parts. Deciduous trees begin to shed their leaves, having lost their green pigment to reveal reds, yellows, oranges and browns, whilst many ripe fruits and crops are also harvested at this time. Harvest celebrations have dominated this season since ancient times, and persist in harvest festivals and Thanksgiving to this day.

Winter

Winter begins in the Northern Hemisphere on about 21 December, and in the Southern Hemisphere around 21 June, lasting until about 20 March and 20 September respectively.

As in autumn, the Earth is tilted away from the sun, and now days are at their shortest and temperatures at their coldest. The season is characterized by high levels of precipitation, and in northern latitudes particularly, snow, blizzards and freezing fog. Plants may die completely, or die back during winter to grow again in spring, whilst many birds migrate to warmer climes, and animals may hibernate. Others, however, may remain active, living off food stored during the summer and autumn months. Many mammals grow thicker coats to insulate themselves, and in areas that often experience snow, birds and mammals may adopt white fur or plumage for camouflage. People may enjoy a range of winter activities at this time of year, such as sledging and ice-skating, and the Christmas celebration falls just after the winter solstice.

Above right: In winter the snow settles on mountain areas, and people can enjoy skiing and snowboarding.

Right: Winter sky with powerlines. The winter solstice, 21 December, occurs on the shortest day of the year. It marks the beginning of winter in the Northern Hemisphere.

The Earth as a System

The Earth can be considered as a system in which all components interrelate. These components are the atmosphere (air), lithosphere (land), cryosphere (ice), biosphere (plants and animals) and hydrosphere (water).

The system begins with the inflow of energy from the sun. It ends with the outward flow of energy back into space, so preventing the Earth from overheating. In between these inward and outward flows the energy warms the planet, drives the atmospheric engine and drives the water cycle. The latter is the process by which water changes between liquid, solid and gaseous forms. Some other chemicals, such as carbon dioxide, also move through the cycle.

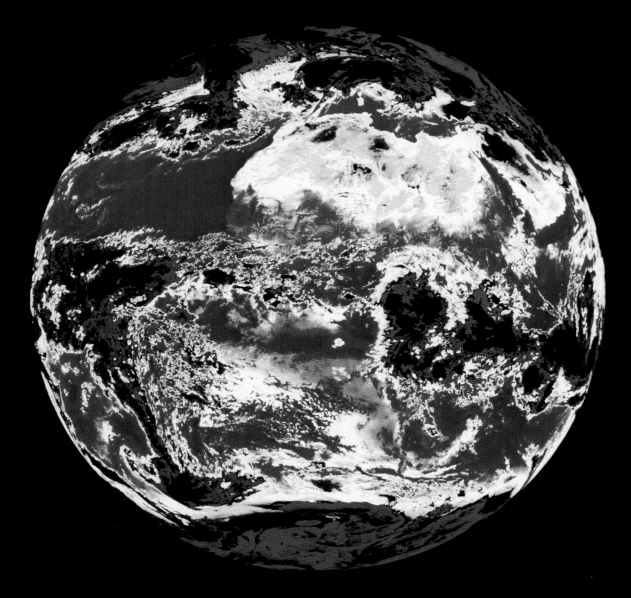

Above: Colour-coded infrared view of the Earth made by a Meteosat weather satellite. The colours represent temperature, from blue (coldest) to red (warmest). Blue features are tops of large cloud banks. Lower, and thus warmer, clouds are shown in yellow and pink, particularly across the centre of the frame. Red areas are generally seas and oceans seen through breaks in the cloud. Land areas, such as the Sahara Desert and North Africa (yellow/green, above centre), are often cooler than surrounding seas during the night as a result of radiative heat loss.

The Energy Cycle

The sun provides the energy which drives the atmosphere, affects temperatures and so creates our weather.

Incoming solar radiation – energy – enters the Earth's atmosphere. Some is able to reach the Earth's surface, heating it. The warmed Earth in turn heats the atmosphere above it. Some energy is absorbed and reflected by the gases in the atmosphere. Some energy returns to space. Some is used to evaporate water which subsequently condenses and cools to form clouds. These in turn absorb and reflect more energy.

It is the constant recycling of the solar radiation received that fundamentally causes our weather.

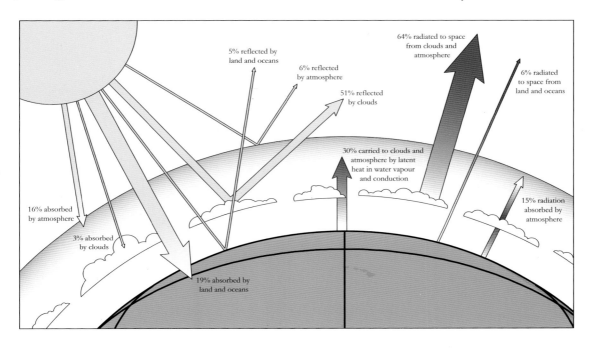

Above: Radiation from the sun reaches the ground and is then absorbed by the atmosphere and surface of the Earth. A proportion of this radiation is reflected back into space. Gases such as carbon dioxide trap some of this reflected radiation, with the result that the average surface temperature increases. This is known as the greenhouse effect because the layer of carbon dioxide has a similar effect to the glass in a greenhouse, which lets heat in and traps it inside.

Right: A satellite map of the world's average surface temperature. Colours represent temperature as follows: mauve = −38°C/−36°F; blue = −36 to −12°C/−33 to 10°F; green = −10 to 0°C/14 to 32°F; yellow = 2 to 4°C/36 to 39°F; pink and red = 16 to 34°C/61 to 93°F; deep red and black (as in Australia) = 36 to 40°C/97 to 104°F. The data for the map was recorded by the High Resolution Infrared Sounder (HIRS) and Microwave Sounding Unit (MSU) instruments on board a satellite of the US National Oceanic and Atmospheric Organization (NOAA).

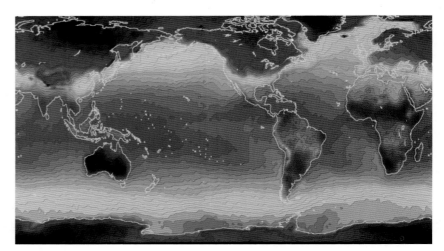

What is the Atmosphere Like?

The atmosphere around our planet is critically important for our survival. It absorbs the energy we need, reflects the surplus back into space and provides a layer of protection against the harmful elements of incoming radiation.

It contains gases, clouds, particles of dust and other particles called aerosols. The main gases are nitrogen and oxygen.

It is arranged in distinctive layers with no clearly discernible outer limit. However, most of the atmospheric mass lies below an altitude of 100 kilometres /62 miles. It is very thin; an analogy is that the atmosphere is to the Earth the equivalent of a layer of skin to an onion.

In recent years concern has grown at the potentially harmful changes believed to be taking place in the gaseous composition of the atmosphere.

Right: Noctilucent clouds, seen from the International Space Station (ISS). These clouds (white streaks across centre) are very thin and form high above the Earth at heights of 75–90 kilometres/47–56 miles. Because they are so thin and high, they can only be seen at twilight, shining in the glow of the setting sun, and only at high and low latitudes. The lower three layers of the atmosphere (from bottom) are the troposphere (orange, 15 kilometres/9 miles thick), and the stratosphere and mesosphere (both pale blue and 35 kilometres/28 miles thick). Above these, the atmosphere fades into the darkness of space. A crescent moon is also seen. The ISS orbits some 380 kilometres/236 miles high.

Exosphere (690–800 km)

Thermosphere (80–690 km)

Mesosphere (50–80 km)

Stratosphere (18–50 km)

Troposphere (0–18 km)

The red line depicts the
Ozone Layer

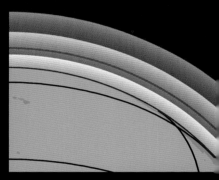

Above: Scientists have divided the Earth's
atmosphere into four zones based principally on
their temperature, density and chemical
composition. The troposphere, closest to the
surface, is the densest part of the Earth's
atmosphere. The stratosphere is less dense and is
also warmer because it absorbs much of the sun's
radiation, thus providing a protective blanket for the
Earth. The mesosphere is cooler than the
stratosphere, but temperatures in the
thermosphere rise dramatically with proximity to
sun and reach around 1700°C/3100°F.

WIND

Global Wind Patterns

It was Aristotle who first commented in *Meteorologica* that global wind patterns were caused by the heating of the atmosphere by the sun.

Because the Earth's surface heats up unevenly there are varying bands of airflow which produce different weather at different locations.

Between the Tropics there is relatively even, intense heating. This produces powerful convection currents in the air. When air rises the resulting pressure at ground level is lowered. As the air here is warm, it rises, hence the low pressure. This occurs in a belt all around the globe. The air that has risen eventually reaches the troposphere, can go no further and so cools and descends in the region between 30°

N and S. The air that has risen is replaced by air moving from about 30° N and S. The resulting circulation is known as a Hadley cell, after the 18th-century scientist who first suggested the idea. These cells give rise to the trade winds and carry warm air away from the Equator. The area at the Equator, where there is little wind, is the doldrums. At lower latitudes air continues to rise and fall, towards the poles, in a circulatory motion. These are known as Ferrel cells. These carry cool air towards the Equator. The theoretical movement of winds should be N–S but in reality they move in a generally NE direction. This is due to the Coriolis force set up at the Earth's surface as it spins on its axis.

Above these cells and somewhat separate from them are the jet stream winds.

The Coriolis effect

The Coriolis effect is caused by the rotation of the Earth. The effect deflects objects moving along the surface of the Earth to the right in the northern Hemisphere and to the left in the southern hemisphere. As a consequence, winds around the centre of a cyclone rotate anticlockwise in the northern hemisphere and clockwise in the southern hemisphere.

If the Earth's surface were smooth, uniform, and stationary, atmospheric circulation would be very simple. Unequal heating is the main driving mechanism responsible for the Earth's atmospheric circulations and gives rise to the three cell circulation model: Hadley, Ferrel and Polar cells.

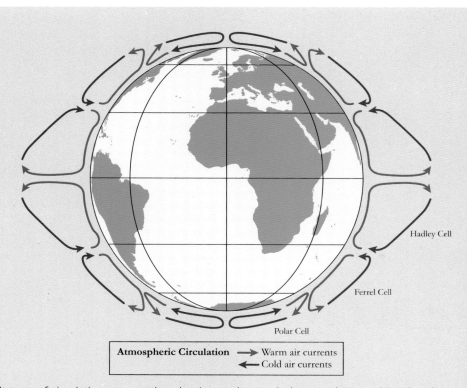

Hadley Cell

Ferrel Cell

Polar Cell

| Atmospheric Circulation | → Warm air currents |
| | ← Cold air currents |

The Hadley cell is a simple conveyor belt type of circulation pattern that dominates the tropical atmosphere and is related to the trade winds, tropical rainbelts, subtropical deserts and the jet streams. The Ferrel cell describes the surface flow in the temperate zone of air that becomes the Westerlies (prevailing winds between 30° and 60° latitude). The Polar cell is the northernmost cell of circulation between 60°N and the North Pole and produces winds that are often weak and irregular.

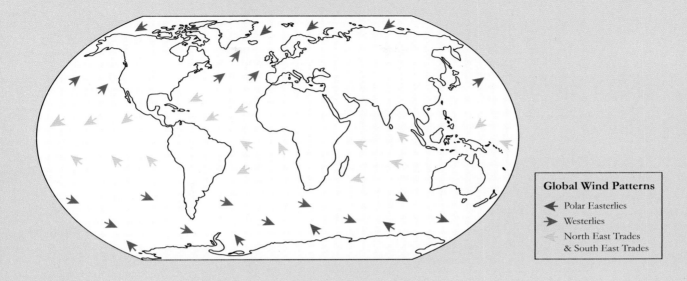

Global Wind Patterns

◀ Polar Easterlies

▶ Westerlies

◀ North East Trades
& South East Trades

Below: Summer winds contribute to the spread of fires in areas where conditions are dry.

Jet Stream

The upper atmosphere contains large, vertical cells of wind which rotate. These help to redistribute heat away from the Equator. They are usually stable but occasionally break up and re-form and have been associated with unusual weather such as drought and flood. The major British drought of 1976 is believed to have been caused by the persistence of high pressure over the country. The normal movement of low pressure is believed to have been dragged north by the jet stream.

Air temperature decreases from the Equator to the poles and wind speed in the upper air is proportional to the rate at which temperature changes. This is most extreme where the equatorial and subtropical air masses meet, which is at the polar front, and where equatorial air meets tropical air. In these areas narrow belts of wind – jet streams – blow, often with great force, between 160–240 kilometres/100–150 miles per hour and sometimes in excess of 320 kilometres/200 miles per hour. It is believed that movement of the jet stream can significantly affect normal weather patterns.

Above: Jet stream clouds seen over north-western Libya. The jet streams are high-altitude, fast-moving air currents that are a few thousand kilometres long. Cirrus clouds form at these high altitudes (these winds form 7–14 kilometres/4–9 miles high) and are swept along in the jet stream. The wind speeds can range from 60–125 kilometres/37–78 miles per hour. The jet streams form as a result of thermal effects, forming at the boundary between a warm and a cold body of air.

Right: Earth's 3-D cloud cover. Coloured composite image of the Earth, showing surface temperature (various colours) and three-cloud cover (white). It was created by integrating multiple satellite data, and is the first representation of the Earth's cloud cover in three dimensions. It combines electromagnetic, spectroscopic and thermal measurements of cloud cover and land and sea temperatures. Temperature is colour-coded, from red (high) through to dark blue (low). The warm water current of El Niño in the Pacific Ocean is clearly seen (red band at centre). Red areas seen on the land masses of North and South America (green/brown) represent forest fires.

Air Masses

Because the Earth does not heat up evenly, some places will be warmer than others. These sections of air can often cover huge areas of the Earth's surface, even as large as entire continents. Such areas of air are called air masses and have more or less the same temperatures, pressure and humidity. These air masses inevitably move, as warm air rises and is replaced by colder air.

An air mass is called continental (dry) or maritime (moist) depending on whether it was formed over the land or sea. It is either polar (cold) or tropical (warm) depending on the latitude where it was formed.

Where air masses of different temperatures meet they form fronts, imaginary lines drawn on weather maps, which are more properly called synoptic charts. This situation is typical of that which causes depressions, low pressure systems, to form.

Right: Northern Cape, South Africa. In summer, much of southern Africa is affected by maritime tropical air which is characterized by high temperatures, high humidity of the lower layers over the oceans and stable stratification. Since the air is warm and moist near the surface, stratiform cloud commonly develops as the air moves polewards from its source.

AIR PRESSURE

Global Pressure Patterns

Atmospheric or air pressure is the weight of the air on the land. It is measured in millibars using an instrument called a barometer. The barometric pressure is recorded both locally and globally and when places of the same pressure are joined together with lines called isobars and shown on a map, a complex pattern emerges. Distinct areas of high and low pressure appear which, on a global scale, appear in well-defined bands which largely coincide with latitude.

When air is warmed by the Earth's surface it rises, reducing the pressure on the land surface. Conversely, air at altitude is cooler than that below it so descends and increases the pressure on the Earth's surface. Air naturally moves from high to low pressure. Imagine a toy balloon filled with air. The air inside is under considerable pressure as it is trapped inside the rubber skin of the balloon. Undo the knot at the end and the air inside rushes out towards the area of lower pressure outside. The movement of air from high to low pressure is, of course, wind.

High and low pressure areas are associated with distinctive weather. High pressure creates dry and calm weather which can produce periods of drought or, in winter, frosty conditions. In contrast, low pressure is associated with cloudy, windy and wet weather. Areas of extreme low pressure may develop into tropical storms called hurricanes or typhoons.

Given the exceptionally close association of air pressure with weather, pressure systems are closely monitored by forecasters.

Right: A barograph, a barometer which can measure and record the atmospheric pressure as part of weather forecasting. This is an aneroid barometer in which changes in atmospheric pressure causes a bellows-like vacuum chamber (at centre) to expand or contract. This deformation causes a pointer to move around the face of the graph paper to reveal the atmospheric pressure. The barometer here has been calibrated in units of inches of mercury. High pressure generally indicates a period of dry weather, whilst low pressure weather systems tend to be rainy.

Left: Satellite image of low-pressure off Australia's southern coast. Land is brown, water is dark blue and clouds are white. These systems develop when a pressure difference between the core and its surroundings, combined with the Coriolis effect, causes air to circulate. The resulting swirling clouds, seen at centre, rotate in a clockwise direction in the southern hemisphere.

Below left: Isobaric map of western Europe (green) showing a depression (low pressure) over Great Britain, surrounded by several anticyclones (high pressure systems). The map consists of lines of equal air pressure (isobars). The air pressure in the depression falls towards the centre of its tightly spaced isobars. The air pressure in the anticyclones rises towards the centre of their widely-spaced isobars. They bring settled dry weather with light winds. Isobaric maps are needed for accurate weather forecasting.

Below right: Diagram showing the symbols for cold, warm and occluded fronts.

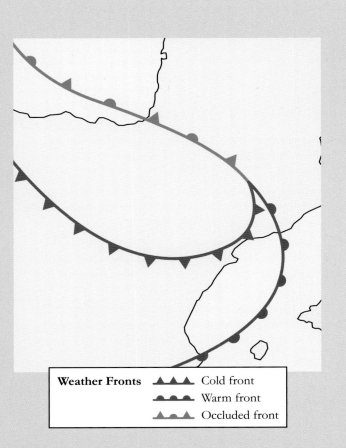

Weather Fronts
▲▲▲ Cold front
●●● Warm front
▲●▲ Occluded front

High Pressure

Air pressure means the weight of the air on the Earth's surface. If the air pressure is high it means that cold air must be descending and adding weight – pressure – on the ground. This air then warms up as it reaches the lower levels of the atmosphere; the air molecules expand, becoming larger than water vapour molecules, which are then absorbed. This means the air is dry – no clouds. The air is calm or has light winds but what wind does exist will blow in a clockwise direction in the Northern Hemisphere and the reverse in the Southern Hemisphere. High pressure systems are often known as anticyclones.

Distinct weather occurs under high pressure conditions. In the summer the characteristic light winds or calm conditions and lack of cloud means the weather is dry and sunny, ideal summer weather. The lack of rainfall, however attractive it may be for holiday-makers and for farmers at harvest time, can be a problem as drought is a characteristic feature of persistent high pressure. Such a situation existed in western Europe during the summers of 1976, 1993, 2003 and 2006 as a result of persistent high pressure settling over the region for as much as three virtually unbroken months at a time.

In the winter the weather conditions are broadly the same as in the summer, dry, calm and with clear skies. A bright, crisp winter's day may be ideal for a country walk but there is a sting in the tail of such weather. The lack of cloud cover creates exceptionally cold temperatures, particularly at night. Any heat generated during the day radiates out into the atmosphere and, with no cloud to trap the heat, frost is likely. If conditions are favourable, fog may also develop, sometimes freezing fog. These are the conditions hated by drivers of cars and other road vehicles. The incidence of road accidents is likely to be higher under these weather conditions.

Right: Satellite image of Europe, with north at the top. The image approximates to true colours with land appearing green, water blue and clouds white/pink. Britain and Ireland lie off the coast of France (lower right) with Spain at bottom centre. Swirling clouds mark the position of a depression or cyclone, an area of low atmospheric pressure, in the Atlantic Ocean at upper left. A cloudless area of high pressure, an anticyclone, lies over the British Isles.

Left: Namaqualand, South Africa. When high pressure lies overhead, conditions tend to be quiescent with light and variable winds. Locations close to mountain ranges often experience variable conditions over a relatively short time because mountains cause air to rise, enhancing cloud formation and precipitation.

Below: High pressure in winter brings blue skies but the lack of cloud cover means that overnight temperatures are low.

Low Pressure

Wet, windy, often cold and dull weather is inevitably the result of low atmospheric pressure systems, or frontal systems as they are often termed. These are common in mid-latitudinal belts around the globe where they may also be known as depressions or cyclones. When seen from space they are identified by the distinctive swirl of cloud that brings the rain and dull weather. In extreme cases the low pressure may intensify to form hurricanes, also known as tropical cyclones or typhoons.

Depressions form where two air masses of different temperatures meet, along an imaginary line called a front that is identified by a swathe of cloud. The cold air is dense and heavy whereas the warm air is lighter. When they meet the warm air mass rises over the cold air, because it is lighter. As it rises it creates an area of low pressure into which the cold air moves; the advancing edge of this is the cold front. The rising warm air creates cloud and, if conditions are right, rain. The fronts begin to rotate upwards and outwards as the air rises because of the Coriolis force, an eddying effect brought about by the Earth's rotation on its axis. The rotation is anti-clockwise in the Northern Hemisphere and the reverse in the Southern Hemisphere.

Three different fronts may exist within the depression depending on the conditions that develop. The cold front marks the advance of the cold air mass while the warm front does the same for the warm air mass. An occluded front occurs when the cold front catches up with the warm front and they become uplifted. This leaves more uniform air at the surface and eventually leads to the decay of the system.

Once formed, depressions are carried east in the Northern Hemisphere along the Rossby Wave system.

Above: False-colour satellite image of a frontal system over the North Atlantic Ocean. The UK and France may be seen at bottom right. At top centre is the ice cap over Greenland. The frontal system is the swirling mass of clouds to the left of centre, with a low pressure area at the centre of the spiral. Low-level clouds are shown as yellow, high-level clouds as white. The pale yellow area at top right is a large bank of sea fog.

Right: Storm clouds seen from the Challenger Shuttle in space. This photograph shows the clouds associated with a cyclonic disturbance. This is found near a low pressure region, and used to be called a mid-latitude cyclone. If the depression at the centre of the storm has a very low pressure, it may lead to relatively violent storms with high winds. The centre of this system is seen as the spiral of clouds just below the Earth's limb near top centre.

Left: Rain squalls blowing over Loch Broom in Scotland. In an area of low pressure the air has a tendency to rise. This upward motion means that there is less pressure from the air pushing down on the Earth. As air rises it cools, and if there is enough water vapour, it may condense to form clouds and rain. This is why low pressure is generally associated with wet weather.

Left: Coloured satellite image of a low pressure frontal system over the North Atlantic Ocean. The system is the swirling mass of clouds (at centre). Low-level clouds are shown as yellow and red, high-level clouds as white. Lines of longitude and latitude are superimposed on the image, as are coastlines. Great Britain, France, Spain and Portugal may be seen at lower right. At top centre are Iceland and Greenland. In a low pressure system the air pressure decreases towards the centre of the spiral of clouds.

Fronts

A front is an imaginary line drawn on a weather map — more appropriately called a synoptic chart — marking the point where a cold air mass meets a warm air mass. Fronts form an important part of a low pressure system or depression.

There are three different types of fronts — cold, warm and occluded. Each one is represented on a synoptic chart as a solid line with the different types of front having a special symbol added to identify them. A warm front has filled semi-circles whereas the cold front has filled triangular barbs. An occluded front is shown as a mixture of cold and warm fronts and has both semi-circles and triangles alternating along the line.

In a cold front dense cold air moves beneath warm air, forcing it to rise. The lifting of the warm air causes cooling, condensation and cloud formation and may produce showers, perhaps thunderstorms.

At a warm front warm air rises up and over a cooler, denser air mass. This forms an area of thickening cloud which gradually reaches lower altitude creating dull, wet conditions.

Occluded fronts occur when a cold air mass catches up with a slower-moving warm air mass. Over the course of about 48 hours the cold air behind the front, as it is dense, pushes beneath the warm air, catches up with the warm front and forces the warm air off the surface. The effects of the warm front are lost and any rain that forms is unlikely to be heavy.

Left: View of an approaching dry squall line. The line of clouds is associated with a cold front and marks the boundary between two air masses of differing temperature. As the squall line passes overhead there is a sudden rise in wind speed, often associated with other violent changes in weather conditions. A veer in wind direction, a rise in pressure and a drop in temperature accompany the passage of the front.

Above: Rain and mist in a wood in autumn.

Right: Space shuttle image of a cyclone from space. A cyclone is a system of winds rotating inwards to an area of low barometric pressure. The cyclone forms the swirl of clouds surrounding the 'eye' (at centre). Cyclones can form into hurricanes or typhoons. The 'eye' comprises relatively calm, cloud-free air at a very low pressure. Around this 'eye' are the cyclone's strongest winds. Cyclones form in tropical regions and are caused by large-scale evaporation and convection due to warm sea temperatures.

Opposite: A rain squall engulfs Gravina Island, Ketchikan, Alaska.

What Happens During a Depression

A depression or low pressure system is formed when a cold air mass meets a warm air mass. These are common in mid-latitudinal areas both north and south of the Equator. They are responsible for a great deal of the annual precipitation that falls across these areas and can produce very wet and windy weather.

A typical depression will take about 48 hours to develop and finally disperse, before re-forming later – a process called cyclogenesis.

The weather that is typical of a depression will vary depending on which particular air masses are involved but will have common characteristics.

As the warm front advances air pressure gradually falls, temperatures remain steady and wind speed increases, backing slightly. Conditions become increasingly overcast and so visibility falls as cirrus clouds are replaced with alto- then nimbostratus clouds. As the front moves overhead the winds veer, temperature and humidity rise markedly and the earlier continuous, heavy rain eases or stops. At this point the area behind the warm front, but ahead of the cold front, is known as the warm sector.

As the cold front advances air pressure falls a little and wind speeds increase but temperatures fall significantly. Cloud cover is low, stratus and altostratus, making visibility poor and bringing light rain which increases markedly as the cold front passes overhead; hail and thunderstorms could occur.

Once the cold front has passed air pressure gradually rises; wind veers and decreases in speed. Temperatures remain low, visibility improves and shower clouds replace the low, heavy, dull conditions; bright intervals interspersed with showers now dominate weather conditions.

Sea and Land Breezes

Anyone who is interested in sailing or windsurfing will be aware that coastal areas experience distinct reversals in wind direction during the day and overnight. This is brought about by the diurnal heating and cooling of the Earth. As the surface heats up it causes thermals to form as the warmed air rises, and stronger upper-level winds are brought down to the surface to replace it. Overnight the Earth's surface rapidly loses heat, most notably if the skies are clear with no insulation from clouds. Temperature inversions may form, where temperatures at altitude are higher than those at ground level, and winds below the inversion become light and variable. Soon after the sun rises the ground is heated and so the cycle repeats itself. Sea breezes are a common feature of coastal locations and they form because of this variability of heating and the resulting air movements – winds.

During the day the land heats up faster than the sea. Air pressure falls over the land as the air rises and cooler air rushes in from the sea to replace it, and a wind forms – sea breeze – which normally reaches its peak in the late afternoon.

Overnight, however, the land cools more quickly than the sea. The surface air, which is cooler than that over the sea, sinks and drains off the land and out over the ocean.

Opposite: Waves break over the Tevennec Lighthouse, France.

Right: As the land heats up in the daytime, the wind comes from the sea (top), but at night the sea retains heat longer so the wind changes direction (bottom).

Below: Curling waves like this are loved by surfers.

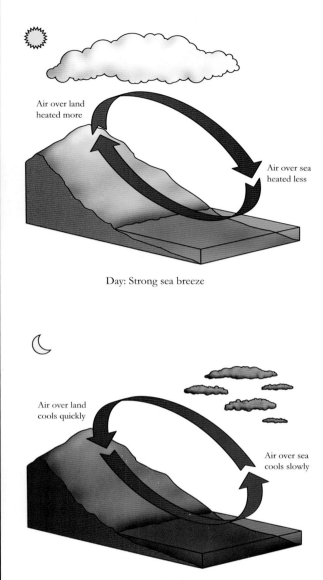

Air over land heated more

Air over sea heated less

Day: Strong sea breeze

Air over land cools quickly

Air over sea cools slowly

Night: Weak land breeze

Monsoon Pressure Systems

Monsoons are seasonal changes in wind, usually bringing heavy rain, that are experienced in many of the world's tropical regions such as India, Bangladesh and Pakistan. A monsoon is a land and sea breeze on an enormous scale, which is produced by seasonal changes in atmospheric pressure systems. During the winter in the Northern Hemisphere dry, sinking air produces the intense high pressure system over Siberia, which creates strong winds which blow away from the sea in a northeasterly direction.

In the summer the land heats up faster than the sea. The high pressure system weakens and a low pressure system develops instead. This draws in southwesterly winds off the sea and produces exceptionally high levels of rainfall.

In Cherranpunji in India the average rainfall in January is 18 millimetres/¾ inch but in July the rainfall peaks at 2446 millimetres/96 inches.

Whilst undoubtedly essential for the people of the area, the rain that falls during the monsoon of the Indian sub-continent can cause numerous problems. Flooding is an annual hazard and can cause severe if short-term disruption to schools, transport and industry.

Although the monsoon is an annual occurrence it can occasionally fail or the rain may be sporadic or lighter than normal. When this happens the consequences can be catastrophic for the farmers whose crops, and therefore livelihoods, depend on the rain that falls.

Large parts of Africa and North and South America also experience less intense monsoons.

Right: Monsoon flooding on the plains of India, showing half-submerged field systems. Monsoon winds blow in India from June to mid-December, but some years they arrive weeks late, and occasionally fail to appear at all. Asia has the most pronounced monsoon conditions – particularly in India, Bangladesh, Burma, Thailand and the Philippines.

Opposite above: Street during heavy monsoon rain in Benares, India.

Opposite below: Diagram of monsoon winds in summer (left) and winter (right) in southern Asia. In summer, the land heats up much faster than the ocean, and the hot air rises, causing a large area of low pressure. This draws in cooler, moister air from the ocean, producing a steady wind from the southwest. Clouds form as the moist air hits the Himalaya mountain range, and these produce a huge amount of rain, which is vital for crops but often leads to flooding. In winter the situation is reversed. The sea cools more slowly than the land so the low forms over the ocean, drawing cool air from the mountains and producing a wind from the northeast.

Below: Pedestrians walk through monsoon rains in Mogadishu, Somalia.

WATER

The Water Cycle

The relationship between water in all its forms and the way it is transferred from one state or place to another is the water or hydrological cycle. It is the basic mechanism by which water is moved. It exists because air absorbs and releases water vapour.

Almost 90 per cent of all the water on Earth is stored in the oceans, rivers and lakes – the hydrosphere. After that the largest store of water is in the form of ice in the world's glaciers or ice sheets – the cryosphere. Sea water comprises over 97 per cent of all water on Earth; the only fresh water on Earth is the remaining three per cent. Of that fresh water nearly 80 per cent is stored as ice. The remaining water is either in the lithosphere – underground, stored in aquifers (22 per cent of the fresh water) – or in the atmosphere. Only one per cent of all freshwater occurs in the form of water vapour, soil moisture, rivers, lakes and inland seas.

The cycle can begin with evaporation from a surface such as the sea. Onshore winds will force the resulting water vapour to rise and if the temperature drops enough it might reach dew point – saturation point. Condensation of water vapour into clouds will then occur with some precipitation resulting. When it reaches the ground a number of different things may happen.

Some of the precipitation will infiltrate into the underlying rocks and could be stored as ground water in aquifers if the right geology – porous rock such as chalk – is present. Some will be absorbed by the soil and may flow through under the influence of gravity to reach the sea once more. Some may fall as snow and over time develop into glacial ice to be largely stored until a future date, when evaporation and melting takes place, changing the ice back into a liquid. Some may be stored in rivers, lakes and inland seas, from which more evaporation can take place. Vegetation is another way in which falling precipitation may be used within the cycle; plants take up the water for growth and transpire the surplus from leaves. Alternatively they may intercept precipitation on their leaves, which is later evaporated and returns to the water cycle.

The water in the seas is of particular global meteorological importance since the oceanic currents redistribute the warm or cold water through the so-called 'great conveyor belt' which is believed to drive the world's weather engine.

Opposite below: The coast near Arniston, South Western Cape, South Africa. Ocean currents have a huge impact on the weather experienced over nearby lands. The air above cold currents carries little moisture, which helps to create deserts on the west coasts of continents. The contrast between cold ocean temperatures and the warm land also produces sea breezes along these coasts.

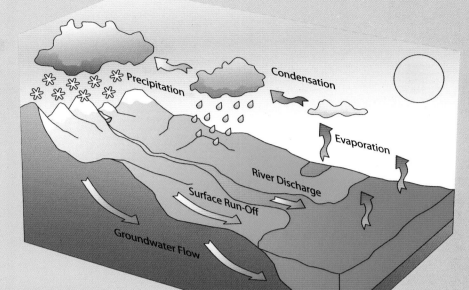

Above: Overview of the Earth's water cycle. The Earth's total water supply is estimated to be over 1.3 billion cubic kilometres. It is constantly redistributed by a variety of processes. Water evaporates from the ocean and may fall as precipitation over land (lower left, top). Water re-enters the atmosphere from land by evaporation and transpiration (water loss by plants) (top). Water flows from land to ocean in surface rivers and streams (centre). It also travels through soil and rock as groundwater (right).

Left: Diagram showing the distribution of water on Earth.

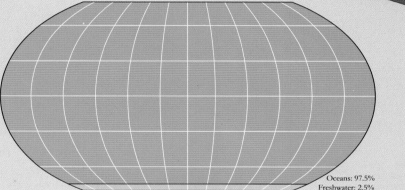

Oceans: 97.5%
Freshwater: 2.5%

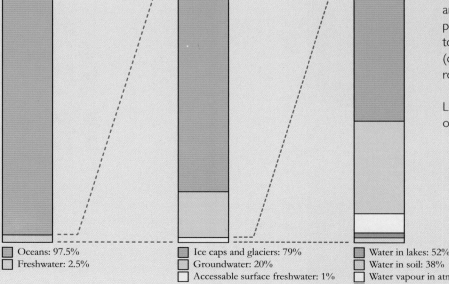

◻ Oceans: 97.5%
◻ Freshwater: 2.5%

◻ Ice caps and glaciers: 79%
◻ Groundwater: 20%
◻ Accessible surface freshwater: 1%

◻ Water in lakes: 52%
◻ Water in soil: 38%
◻ Water vapour in atmosphere: 8%
◻ Riverwater: 1%
◻ Water in living organisms: 1%

Distribution of Water on Earth

Humidity

Humidity is the amount of water vapour, a colourless gas, in existence in the atmosphere. A high humidity means there is a lot of water in the atmosphere. Under these conditions, such as are sometimes felt before a thunderstorm, it can feel 'sticky' or 'close'. A low humidity usually feels more comfortable, but when the air is too dry it can also feel uncomfortable, particularly for asthma sufferers. How humid the atmosphere is depends on where the air has come from and its temperature, and so it will vary considerably between places. At the poles there is no humidity but in the tropics it can be as much as four per cent by volume.

As the air temperature increases water molecules evaporate, some eventually condensing to form a liquid. When equilibrium develops between the evaporation rate and the condensation rate saturation or dew point is reached and so the air can hold no more water vapour. At this point deposition of water takes place on surfaces or in the atmosphere.

There are three ways to measure humidity. Relative humidity is the amount of water vapour as a percentage of the maximum amount a given parcel of air can contain. This is the most commonly used and is often referred to in weather forecasts. It is important as it suggests the likelihood of rain, dew or fog. Relative humidity makes it feel hotter outside in summer as it prevents sweating from the skin, a process which the body uses to cool itself when hot. Absolute humidity is the mass of water in the atmosphere and is expressed as the number of kilograms of water per cubic metre of air. Specific humidity is the ratio of water vapour to dry air in a particular volume of air and is expressed as a ratio of kilograms of water vapour per kilograms of air.

Humidity is measured using an instrument called a hygrometer. It measures both the dry air temperature and the moist air temperature. The relative difference between them is a means by which it is possible to calculate relative humidity.

Water vapour is another greenhouse gas, one that absorbs energy that would otherwise escape into space and so causing the earth's temperature to increase. Interestingly, when the air warms it has an increased ability to hold water vapour and so more heating takes place and so on.

Left: Rainforest has a very high annual precipitation level with a short rainy season, thus humidity is high. However, the amount of water vapour carried in the air is further raised by the fact that much of the evaporation cannot rise above the tree canopy to be blow away. This vapour therefore falls again as rain.

Opposite above: The waterfall at Victoria Falls, Devil's Cataray, forms part of the Zanbezi River which divides Zambia and Zimbabwe. Zimbabwe has a tropical climate, moderated by altitude, with a rainy season between November and March.

Opposite below: Aqua satellite map of water vapour in the atmosphere over the oceans. The amount of water vapour (which forms clouds) is colour-coded, from light blue (least) to dark blue (most). The small yellow areas correspond to areas of high precipitation. Snow and ice are yellow and deserts are dark green. Other land is black.

Above: Hoar frost forms when air saturated with water vapour cools rapidly and the vapour precipitates out directly onto a freezing surface. A layer of ice crystals, often with intricate flowery patterns, then forms on the ground, plants, windows and other objects.

Right: Icicles covering new buds on a fruit tree. An icicle is formed when above-freezing water runs or drips into sub-freezing air. The water freezes as it moves, forming a narrow cone pointed downward, which grows in both length and width, being widest at its top.

Opposite: Dew on a spider's web. Dew is the term for small droplets of water that appear on objects in the morning or evening and is a form of condensation. Dewdrops are small — less than 1millimetre/$\frac{1}{32}$ inch in diameter.

Dew, Frost and Rime

The dew that appears on the lawn and the frost that appears on car windscreens on winter mornings is the result of air close to the Earth's surface being saturated with water vapour. During the day the ground is heated by incoming solar radiation which in turn heats the air. At night, however, the ground cools rapidly, particularly if the sky is clear. The air in contact with the ground cools but the air above is warmer (air is a good insulator). The air at or near the ground cools to saturation at the dew point, the temperature at which water vapour starts to condense. When the saturation point is above the freezing point of water (0°C/32°F) dew will form but if it is below freezing point it is frost that will form.

When the forecasters refer to a ground frost it is one in which the ground is frosty but the air temperature is still at three or four degrees. An air frost is one where temperatures in a Stevenson's Screen reach zero even though the ground sometimes remains above freezing point. This is common in autumn as the ground retains a little warmth from the summer.

There are two types of frost, both of which create beautiful features in the landscape, particularly on vegetation. Hoar frost forms when the air cools and water condenses onto the grass in a delicate icy structure. It only forms when winds are very light and creates a definite, delicate crystalline structure growing on elements in steps or layers. The white colour is caused by small air bubbles trapped in the ice which reduce its transparency. The smooth faces of the hoar crystals allow it to glitter in the sunlight, particularly at the low sun angles of early morning. Rime occurs when frost formation occurs quickly and is common during fogs where near-freezing water droplets come in contact with freezing surfaces On close inspection it has a grainy appearance, like sugar or salt, forming spikes, needles, or feathers and with no recognizable crystal structure, and is denser and harder than hoar frost.

Wind speed can affect the severity of frost. When winds are strong they slow down night cooling and reduce the risk of frost. If winds are strong after the temperature has fallen below 0°C/3C°F it can create damaging and penetrating frost.

It is important for a number of people to have an accurate forecast of frost. Farmers need to know when crops are at risk and drivers and motoring organizations must be aware of icy patches on the roads. Local governments need to spend considerable sums on gritting and salting roads to prevent accidents.

CLOUDS

What is a Cloud?

A cloud is a collection of water droplets or ice crystals that are packed together densely enough to make it visible. Water vapour needs a nucleus in the atmosphere for condensation or accumulation to take place. This is usually something in the atmosphere such as dust and is called a hygroscopic or condensation nucleus.

Right: Lenticular cloud. Fish-eye view of a lenticular cloud (altocumulus lenticularis) over Sandstone Bridge, Alabama Hills, California, at sunset. A lenticular cloud gets its name from its smooth, lens-like shape. When wind blows across mountains it tends to cause waves on the mountain's lee side. Lenticular clouds form when moisture in the air condenses at the top of these waves. If the mountain has a regular shape and the wind blows at a constant speed, the cloud pattern will be stable, have a regular shape and remain virtually stationary in the sky for long periods. The Sandstone Bridge is a natural arch formed by weather erosion.

How are Clouds Formed?

Clouds form where water vapour is cooled to saturation – the dew point. This is then followed by condensation, leading to the formation of water vapour droplets or deposition of ice crystals. Temperature falls with altitude – the lapse rate. At high altitude all clouds are made of ice crystals but at low altitude they are composed of water droplets.

Clouds can be associated with convection, mechanical (orographic) or dynamic uplift. Convection means that warm air becomes increasingly buoyant compared to the surrounding air and rises. Mechanical or orographic uplift occurs when air moves over mountains and dynamic uplift refers to large-scale air movement into low pressure systems or along fronts. Essentially, clouds form where there is uplift, cooling and condensation.

Below: Cumulonimbus clouds in the sky over water. Cumulonimbus are large, turbulent clouds which carry rain or hail.

Opposite above: Salt pan and clouds in Botswana. Cumulus clouds form at low altitude from water droplets that condense out of rising columns of moist air

Opposite below right: Clouds seen over the North Sea. The study of clouds is called nephology.

Above: Warm, moist air rises and as it cools at higher altitudes, water vapour condenses and forms clouds.

Cloud Types

Clouds can be classified according to form and altitude on what is sometimes called a 'cloud atlas'. According to the World Meteorological Organization there are 10 types of cloud. These are classified according to the altitude at which they develop and the form or shape that they take.

There are two basic types of cloud: layer or stratus cloud which covers the sky completely and is associated with stable air conditions; and heap cloud, clumps of cloud, usually cumulus.

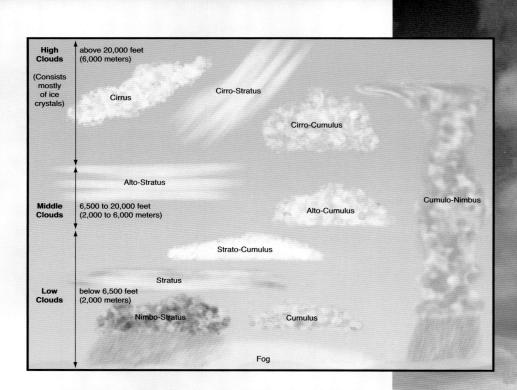

High Clouds (Consists mostly of ice crystals)	above 20,000 feet (6,000 meters)	Cirrus · Cirro-Stratus · Cirro-Cumulus
Middle Clouds	6,500 to 20,000 feet (2,000 to 6,000 meters)	Alto-Stratus · Alto-Cumulus · Cumulo-Nimbus
Low Clouds	below 6,500 feet (2,000 meters)	Strato-Cumulus · Stratus · Nimbo-Stratus · Cumulus · Fog

Above: There are ten basic cloud types.

Right: Spiralling cirrus Kelvin-Helmholtz cloud. Its distinct corkscrew pattern is formed by layers of air moving across each other in opposite directions, creating a series of circular air movements in between them. This type of cloud generally forms at high altitude and dissipates rapidly.

Opposite: Scattered plumes of cirrus cloud in the sky. Depending on their shape, these clouds are also known as 'Mare's Tails' or 'Stringers'. Cirrus is a high-altitude ice crystal cloud which forms delicate filaments or fibres. This cloud type is usually associated with fair weather.

High-altitude Clouds

These are largely composed of ice crystals and their base should be at or above 20 kilometres or 12 miles.

Cirrus
These form at or above 20 kilometres/12 miles but that altitude is lower at the Poles because of the lower temperature. They are formed of ice crystals and have a feathery or wispy form.

Cirrostratus
These are layer clouds which cover most of the sky. They usually herald bad weather and sometimes form a halo around the sun or moon.

Cirrocumulus
These are ice clouds usually with irregular attractive patterns. They are a heaped form of cirrus.

Above: 'Mare's tail' cirrus cloud. Cirrus is a high-altitude cloud, formed of ice crystals. It appears as delicate white filaments, patches or streaks. The long streaks with feathery ends, as seen here, have the common name of 'mare's tail'. Isolated cirrus is sometimes known as an emissary sky, as it often heralds an approaching cyclonic storm.

Oposite above: Cirrus (top), cirrocumulus (in main cloud layer) and cumulus (low on the hills) cloud types visible over the Himalayan foothills of east Nepal.

Right: Cirrus clouds are found at high altitudes, typically above 5000 metres/16,500 feet. They consist of tiny ice crystals, which are blown by the wind into characteristically wispy shapes. A bank of cirrus clouds may indicate an approaching weather front.

Opposite below: Cirrocumulus undulatus clouds at sunset. The pink tinge is due to the reddish light of sunset reflecting off the bottom of the clouds. Their rippled appearance, which occurs due to wind shear, led to their being named undulatus (undulating). If they increase in coverage, they are indicative of an approaching frontal weather system.

Medium-altitude Clouds

These develop at heights of between 2500–5500 metres/8000–18,000 feet.

Altocumulus

These are flattened globes or globules of cloud which are a mixture of ice and super-cooled water. They are white and grey in colour and are often a first sign that thunderstorms will follow.

Altostratus

As the name suggests, these are layer clouds of a dull grey colour and quite often make the sun have a milky appearance. They are often a first sign that rain is to follow.

Nimbostratus

These are thicker, lower versions of altostratus and are always associated with rainfall or snow. They can, however, be so thick and dark that they make the day appear very dark.

Left: Altocumulus clouds forming a mackerel sky. This spectacular pattern may indicate an approaching weather front and impending rain.

Opposite above: Altocumulus clouds lit up by the setting sun. Altocumulus clouds tend to be larger and darker than high-altitude cirrocumulus, and smaller than the low-altitude stratocumulus.

Opposite below: Iridescent altocumulus clouds. When light shines through cloud, it can be refracted to produce an iridescent, multicoloured sky. The iridescence only occurs when sun or moonlight is not directly overhead and the cloud is newly formed. It is a relatively rare phenomenon, occurring mostly over mountains in winter. The altocumulus clouds are composed of water droplets and ice crystals. They occur at medium levels between 2000–7000 metres/6500–23,000 feet.

Low-altitude Clouds

These are water clouds whose bases form at around 2000 metres/6500 feet.

Stratocumulus
These clouds are composed of water and are grey and white with darker areas inside. Their appearance is rounded and rolled.

Stratus
Water clouds which are grey in colour, these form uniform layers which bring drizzle or snow. They are the lowest clouds in the sky and in fact form 'fog' if over hills or coasts.

Cumulus
These heaped clouds are often described as 'cauliflower-like' in appearance. They are grey at the base and white at the top.

Cumulonimbus
These are the classic anvil-headed thunderclouds which can reach as high as 12,000 metres/40,000 feet into the atmosphere. This high in the upper atmosphere, the tops of the clouds are icy.

Opposite: Sky filled with 'fair-weather' cumulus clouds over water at sunset, stretching as far as the horizon. 'Fair-weather' cumulus are small clouds of no great vertical extent and are associated with mild weather. Cumulus clouds have rounded tops and a 'cotton wool' appearance. Their flat bases, as seen here, show a clear condensation level at the level at which the cloud forms. Cumulus may be produced at a frontal system boundary or as a result of thermal cell activity, with a low-altitude cloud base at 300–1000 metres/1000–3300 feet. The depth will generally vary between 500 metres/1600 feet and 1500 metres/5000 feet.

Left: Close up view of cumulus clouds in a blue sky. The word 'cumulus' means 'heap' in Latin, describing the appearance of this thick, puffy type of cloud.

Right: Layer of stratocumulus cloud (white) over the Pacific Ocean, seen from space. These common clouds, at an altitude of 12,000 metres/40,000 feet, do not usually bring rain.

Above: Kalahari sky scene, South Africa. The red, yellow and gold colours arise because the air itself, small dust and aerosol particles scatter short wavelength blue and green rays much more strongly than longer wavelength yellow and red. The remaining direct unscattered light is dimmed, but relatively enriched in reds and yellows. The sunset rays are sometimes reflected back and forth between clouds and the ground. All this goes to make a spectacle painted with every colour and shade of the palette.

Right: Sunset at sea, Northern Natal, South Africa. Sunset rays pass through the lower atmosphere which acts as a giant lens and refracts low sunset rays into long curved paths. Air, dust aerosols and water drops scatter and absorb the rays.

Opposite above: Thunderclouds are large and turbulent clouds carrying rain or hail. When thunderstorms amalgamate, a wide area of land can be overshadowed by cloud.

Opposite below: A cloudy horizon and dust in the atmosphere that scatters light can produce spectacular colours at sunset.

Polar stratospheric cloud
Opposite: Polar stratospheric cloud (white) with other clouds coloured by a sunset. This is a nacreous or mother-of-pearl type of polar stratospheric cloud (PSC). PSCs are high-altitude icy clouds that are found about 24 kilometres/15 miles up in the polar stratosphere. They only form during winter when the temperature drops below around −80°C/−112°F. PSCs play a key role in the destruction of ozone by chlorine. They convert harmless forms of chlorine into highly reactive forms. Chlorine is carried into the stratosphere in CFCs (chloro-fluorocarbons), gases used as refrigerants and during foam plastic manufacture. Photographed over Swedish Lapland in early January.

Unusual Clouds

Lenticular Clouds
As air moves over isolated hills or mountains it is forced to rise to go over them. As it does so the water vapour in it cools to dew point and condenses forming clouds. The clouds that develop are hump-backed or lens shaped, hence lenticular, and normally develop at right angles to the wind direction. Some mountains are almost permanently shrouded by such a cloud, for example peaks in the Himalayas and Table Mountain in Cape Town, South Africa which is frequently covered by the lenticular cloud known locally as 'the tablecloth'.

Mountain Wave
When air blows over a hill or mountain it first has to rise to overcome the obstacle of the hill itself and then it falls down the leeward side. This motion is wave-like in character. Each time the air rises it reaches dew point and cloud formation takes place. The cloud then disperses as the air falls down the leeward side of the hill. Despite the air movement the cloud that has formed appears stationary. This is a form of lenticular clouds.

Nacreous or Mother-of-Pearl
These are very high-altitude clouds, formed at heights of up to 32 kilometres/20 miles in the stratosphere and mesosphere. They are rare, seen only at dawn or dusk when the Sun is about 5° to 10° below the horizon and look similar to high cirrus. At this angle the sun illuminates the clouds to stunning effect, creating beautiful mother-of-pearl-like colours. They do not appear to move and are probably formed as a result of rapid uplift of air over a mountain range.

Noctilucent
These are fast moving clouds which form at extremely high altitude, at about 58 kilometres/36 miles, at the top of the mesosphere and above the height of satellites. They are visible only during the summer between about 50° and 65° north and south and, like nacreous clouds, are seen when the sun is between 6° and 12° below the horizon while the ground and lower layers of the atmosphere are in shadow. They are very fast-moving and, although their formation is still unclear, it is believed they are associated with the jet stream, hence speeds of 480 kph/300 mph.

Contrails and Distrails
The path left by aircraft engines is a contrail, sometimes called a vapour trail – effectively man-made clouds. As the plane flies it releases water vapour and gases that condense rapidly in the cold air. Low pressure can form small eddies at the aircraft wingtips leading to water vapour condensing or freezing, causing smoke-ring-like effects along the contrail.

Lenticular cloud

Opposite: Lenticular cloud or altocumulus lenticularis. Lenticular clouds get their name from their smooth, lens-like shape. They are usually caused when wind forms standing waves of air on the lee side of mountains. Lenticular clouds form when moisture in the air condenses at the top of these waves, and dissipate when the air descends again. The cloud pattern depends upon the wind speed and the shape of the mountains. A constant wind may produce clouds which are stable and remain virtually stationary in the sky for long periods. Lenticular clouds are sometimes mistaken for alien 'flying saucer' spacecraft. Photographed from Mauna Kea, Hawaii, USA.

Right: Aqua satellite image of clouds showing a Von Karmen vortex street (braid-like area, centre) over Jan Mayen Island (at top of street). The wind is blowing from the top. Von Karmen vortices form when low winds blowing over a flat sea are disrupted by an obstacle, such as an island. The vortices are aligned with the wind in a row, or street, with alternate vortices rotating in opposite directions.

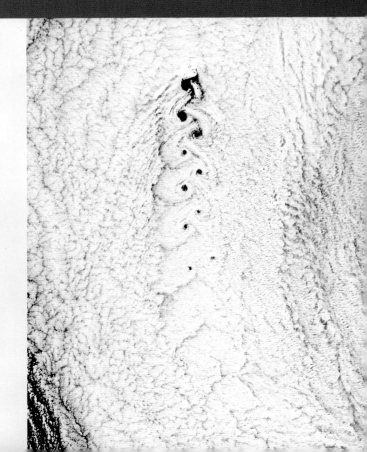

Above: Crepuscular rays (alternating dark and light) are caused by scattered sunlight, as a result of atmospheric particles or a break in a layer of cloud. The divergent appearance is an effect of perspective. The rays are in fact parallel.

Above: Contrails are artificial clouds formed from frozen water droplets from the exhaust of the engines of an aircraft. They form when aircraft are flying at altitudes exceeding 5000 metres/116,500 feet and when the upper atmosphere is nearly saturated with moisture. In dry air the contrail quickly disperses. Sometimes these 'chromosome-shaped' patterns form as the contrail dissipates.

Above: Lenticular cloud photographed in Colorado. As air moves over isolated hills or mountains it is forced to rise to go over them. As it does so the water vapour in it cools to dew point and condenses forming clouds. The clouds that develop are hump-backed or lens shaped, hence lenticular, and normally develop at right angles to the wind direction.

Fog and Mist

Fog and mist are, essentially, clouds formed at ground level. Like clouds, mist and fog are formed when water vapour is cooled to dew point and condenses. Both are made of tiny water droplets in suspension in the air with the difference between them being the density of droplets. The officially accepted definition of fog is when there is visibility of less than 1000 metres/3300 feet. This is used in aviation but a more conventional definition as far as most people are concerned is where visibility is 200 metres/650 feet or less. This happens when there is a high density of water droplets in the area. When visibility falls below 50 metres/165 feet severe disruption to transport occurs with significant risk of accidents.

Radiation fog

This is the most common type or cause of fog. The ground absorbs heat during the day but at night it radiates outwards cooling both itself and the surrounding air. If it cools enough then some water vapour may condense into cloud droplets on the ground and form fog. The ideal conditions for this to form are slight winds, needed to stir up the cold air and create condensation across the wider area, clear skies and long nights. After dawn the incoming solar radiation warms the land and atmosphere and so the fog 'burns off'. In winter the temperatures may remain so low that the fog can last all day.

Above: The rusty red twin towers of the Golden Gate Bridge, San Francisco, rise out from the dense fog. Fog comprises tiny droplets of water. It forms when relatively warm, moist air is cooled. As it cools, some of the water vapour in the air condenses to form water droplets. This is because cool air can carry less water vapour than warm air.

Advection fog

This is formed when a very warm moist air blows over a cold surface. The lower layers of air are cooled down to a level at which condensation occurs, creating fog. This type creates the characteristic fog over the Golden Gate Bridge in San Francisco in California, USA. Moist onshore winds blow across the cold water of the Pacific during the summer months.

Hill fog

This is formed as mild moist air flows up the windward side of a slope where it cools and, if the air is saturated, condenses. If this occurs below the hill top it forms fog.

Coastal fog

This forms when a warm stream of air passes over cold water cooling the lower layers of air below dew point and creating fog. This type can be seen to be literally coming in on the tide.

Freezing fog

This is composed of super-cooled water droplets. These are ones that remain liquid even if the air temperature is below freezing point. A characteristic of freezing fog is rime, the feathery ice crystals that are deposited on the windward side of vertical structures such as telegraph poles and fence posts.

Smog

This is a mixture of smoke and fog. Air pollution creates ideal conditions for the formation of water droplets and when these mix with pollution caused by burning fossil fuels it creates a thick, noxious fog. The infamous London 'pea-soupers' of the 19th and early 20th centuries are typical examples of urban smogs.

Above: Vehicles driving through fog. Fog reduces visibility and although most sea vessels can penetrate fog using radar, road vehicles have to travel slowly and use low-beam headlights. Localized fog is especially dangerous, as drivers can be caught by surprise.

Below: Fog in the North Sea. False-colour satellite image of north-west Europe, showing a large fog bank in the North Sea (pale yellow). The colours of surface features have been processed to approximate to natural tones. The British Isles and the coasts of northern Europe and Scandinavia are seen clearly. Thin, high-level clouds are seen over parts of Germany and Austria (bottom right), partially obscuring the snow-covered Alps. Most of Sweden (top right) is also covered in snow. The reddish tints on the left are due to sunglint from the water of the Atlantic – the Sun was at a low illumination angle at the time of this image made by an NOAA satellite.

PRECIPITATION

Any liquid, or frozen liquid, that falls from the atmosphere to the Earth's surface under the influence of gravity is classed as precipitation. Precipitation is hugely important as it keeps water circulating between the various stores in the atmosphere, oceans, glaciers and ice caps, lakes and rivers.

There are three types of precipitation classified according to their state: liquid (water), solid (ice) or freezing (snow or hail). These are snow, hail, sleet, drizzle and, of course, rain.

Only those clouds that are at least 12,000 metres/4000 feet deep will cause precipitation to occur. Nimbus and cumulonimbus create the heaviest precipitation. Some clouds will cause precipitation but this will not always reach the ground surface as the raindrops may evaporate before reaching the cloud base. The total amount of rain that falls on Earth each year is 5000 million million tonnes. A typical raindrop contains a million times more water than the average cloud droplet and the fastest raindrops can fall at about 9 metres/30 feet per second. In contrast, snow only falls at 0.5 –1 metre /1.5–3 feet per second.

Many parts of the world receive virtually no precipitation of any form. In the northern Atacama Desert of Chile the town of Calama received no rain for 400 years. All that changed, however, in 1972 when torrential rainfall occurred, causing flooding, landslides and a great deal of damage to the town.

Below: Snowfall in New York.

Above: Falling raindrops are often depicted as teardrop-shaped in cartoons or drawings. In fact this is only true when they are dripping from a surface. Small raindrops are spherical. Larger ones become increasingly flattened on the bottom and very large ones are shaped like parachutes.

Right: View of distant rainfall in Death Valley, California, USA. The shower can be seen (lower centre) as a grey/pink sheet of rain emerging from cumulus clouds. Rain is the precipitation of small drops of water. Air masses acquire moisture as they pass over warm bodies of water or wet land masses. The moisture is carried upwards by turbulence and convection. Rainfall results from the cooling and condensation of this water vapour by winds or by being forced to rise over land masses. Rain is extremely infrequent in Death Valley. This rain fell as an effect of the 1998 El Niño event in the Pacific Ocean some distance to the west.

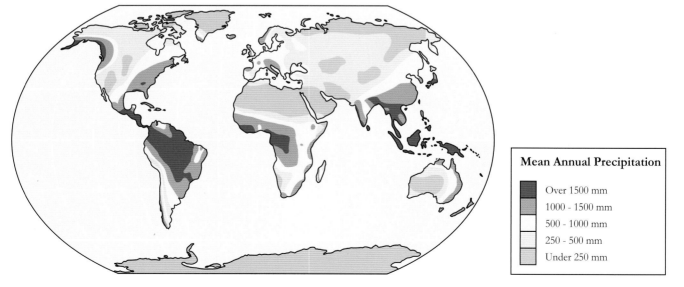

Mean Annual Precipitation

Over 1500 mm
1000 - 1500 mm
500 - 1000 mm
250 - 500 mm
Under 250 mm

Rain

This is the liquid form of precipitation. When rain clouds are sufficiently deep they will produce large drops which fall as rain. This can vary from very light or showery conditions to extremely heavy rain which can lead to flooding.

Drizzle is in the form of a fine mist-like precipitation which appears to drift slowly in the air. Its impact is slight – more an irritation to pedestrians and drivers than anything else – and may not be enough to be measured in a rain gauge.

The process by which rain forms is not simple. The Bergeron–Findelson effect, named after its Swedish and German proponents, is one of the most important theories to explain how and why rain forms. It explains how there is a complex interaction between water vapour, super-cooled water droplets and ice crystals at temperatures between –10°C/14°F and –20°C/–4°F. This helps larger crystals form which eventually fall out of the clouds as snow. When the air temperature is warm enough as the crystals descend the snow melts and turns to rain.

In tropical areas, however, the air temperature is usually too warm for ice crystals to form therefore the rain that falls in these areas is the result of coalescence, in other words the raindrops collide and eventually form large enough drops to fall to Earth as rain.

In thinner cloud conditions the drops will collide with fewer others and so form smaller drops, creating a fine mist – drizzle.

Rain usually comes from layer-type clouds such as stratus but showers are periods of rain between which are periods in which brighter conditions occur.

Rain clouds can form through any one of three main processes. Orographic rain clouds form over hills and mountains while frontal rain clouds form where two air masses meet. Finally, rain clouds also form when convection takes place where warm, buoyant air rises.

Opposite above: Sunshafts burst through rain clouds to illuminate the sea around the Summer Isles, Ross and Cromarty District, Highland Region, Scotland. Raindrops are either water droplets from low-level clouds, or ice crystals from higher-level clouds which melt as they fall through warm air. Raindrops can be up to 5 millimetres/⅜ inch across. Very light rain or drizzle is made up of raindrops less than 1 millimetre/½ inch across. Stratus clouds tend to create long periods of steady rain; cumulus clouds tend to produce short showers.

Below: A veil of rain seen under a large stratocumulus cloud. This sort of veil appears in clouds where there is little convection from below. If the rain evaporates before hitting the ground, it is known as virga.

Snow

For most children – and many adults – snow is the most exciting type of precipitation and is eagerly anticipated as winter develops. Some people still believe that snow is frozen rain but it isn't. Snow forms when tiny ice crystals collide and stick together. Most snow crystals melt as they fall to the Earth but when the air is cold enough they remain and the snowflakes fall as snow.

Each snow crystal has a unique structure and pattern. This depends on temperature, altitude and the saturation of the clouds as they form. They join together when they are wetter and then re-freeze so large flakes tend to form when the air is relatively warm, just above freezing. This situation also tends to develop the heaviest falls as warm air can hold more moisture. Incidentally, this type of snow is the best for those children (and adults) who want to make – and throw – snowballs!

When it is very cold snow crystals tend to be drier and powdery because the ice crystals do not find it easy to stick together. This is the snow that skiers particularly relish and snow forecasts highlight this as ideal for skiing.

The intermediary stage when some of the falling snow has melted but the precipitation lands as half-snow and half-rain is sleet.

The Inuit people have several different words for snow as they recognize, and indeed need to understand the significance of, the numerous types that form.

Right main picture: Heavy snowfall along a creek during a snowstorm in Missouri. Snow is made up of large ice crystals which grow inside clouds from smaller crystals. When it is very cold, snow forms small, dry flakes, making powder snow on the ground. In warmer temperatures larger, wetter flakes are created.

Right: A blizzard is a snow storm normally accompanied by strong winds. It can pile up huge snowdrifts, making it very hard for people or cars to move around.

Right: Snowflakes, light micrograph. In the centre are two plate snowflakes, while around the outside are dendritic snowflakes. Snowflakes are symmetrical ice crystals that form in calm air with temperatures near the freezing point of water. The exact shape of a snowflake depends on local climatic conditions. Snowflakes typically have hexagonal symmetry, as seen here. No two snowflakes are identical, as each experiences a wide range of conditions as it forms inside a cloud.

Hail

Associated with thunder storms and often with tornadoes, hail is an interesting weather phenomenon for most people but for others it is livelihood-threatening and it can kill: 246 people died during a hail storm in Moradabad, northern India, in 1888.

Were you able to slice through a hailstone you would see that it is layered rather like an onion. This layering is a clue to the hailstones beginning as an ice crystal in an unstable cloud such as cumulonimbus or storm cloud and associates hailstorms with violent downpours and storms.

Within the cloud there are strong vertical air currents which throw the ice crystals up and down. and down in the cloud. Eventually the hailstone becomes so heavy that it cannot remain suspended in the cloud and it falls to the ground.

A typical hailstone is around 5 millimetres/³⁄₁₆ inch in diameter but even for one only this size, it is necessary for there to be strong winds – 30 metres/100 feet per second. Although not very common, it is not rare to find golf-ball-sized hailstones falling. Much more rarely considerably larger hailstones form such as the one which fell during a storm in Kansas in 1970. This weighed 765 grammes/27 ounces was 43.68 centimetres/17.2 inches in circumference and almost 15 centimetres/6 inches in diameter. Another even larger hailstone fell in Gopalganj in Bangladesh in 1986. This one weighed in at a hefty 1 kilogram/2½ pounds.

Left: Coloured image of a giant, irregularly shaped hailstone. Hailstones are lumps of ice formed in storm clouds (cumulonimbus) when water droplets remain fluid after their temperature has fallen below freezing point. This super-cooled water circulates in the updraughts within the cloud, forming ice on contact with dust particles in the air. The hailstones grow larger as they circulate around the cloud until they are either too heavy for the cloud to support or are thrown out of the updraught area. Large hailstones like this one (approximately 15 centimetres/ 6 inches across) can damage property and injure people and livestock when they fall.

Above: Shaft of hail pouring from a severe thunderstorm in Gillette, Wyoming, USA. Hail is a form of precipitation made up of balls of ice. The hail forms in cumulonimbus clouds due to violent convection currents. Particles of rain or snow are drawn vertically up through the cloud, and collide with supercooled water and other particles. These collisions lead to the particles sticking together. When they become too heavy to be supported by the air currents, they fall to earth. Some hailstones can be over 10 centimetres/4 inches across, and cause severe damage to property and crops.

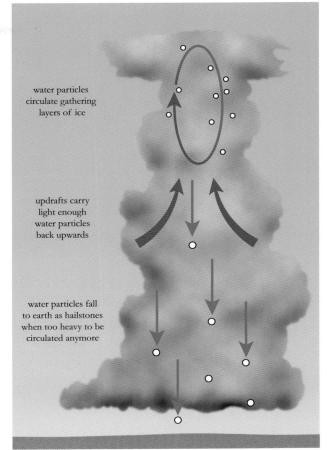

water particles circulate gathering layers of ice

updrafts carry light enough water particles back upwards

water particles fall to earth as hailstones when too heavy to be circulated anymore

Above: Diagram illustrating how hailstones form.

Hailstorms are unwelcome because of the damage they inflict on crops and homes. In the central states of the USA is an area known as Hail Alley where millions of dollars of damage occurs every year when hailstorms arrive. The force of the landing hail flattens crops, often making them unrecoverable. In other parts of the world such as Italy farmers lay out matting roofs on scaffolding poles to help protect vulnerable fruit on trees from the damaging effects of hail. In Sydney, Australia, an area not normally noted for severe hailstorms, one occurred in 1999. In a storm lasting less than 15 minutes, hailstones varying from golf to cricket ball in size fell at speeds of up to 160 kph/100mph severely damaging 25,000 homes, making 44,000 homeless and causing £600 million of damage.

THE ATMOSPHERE: LOCAL PATTERNS

Continentality: The Effects of the Land

The Steppes of Russia and the Prairies of North America conjure up images of huge expanses of grassland. Both areas are far distant from the moderating influence of the sea and so have developed a continental climate, one that is evolved in the middle of a large continent. Such climates are characterized by warm, short summers and long cold winters, often with blizzards. Despite this, continental areas are generally dry: they are a long way from the sea so winds will have lost most of their moisture before reaching the area. They usually have three or more months with temperatures over 10°C/50°F and winters with one month or less with temperatures below 0°C/32°F. During the summer they are dominated by warm air masses and in the summer by cold air masses

This type of climate is caused by the area's distance from the sea. The sea has a high specific heat capacity which means that it takes a long time to heat up and similarly a long time to cool down. The land, however, has a much lower specific heat capacity. When solar radiation reaches the land the soil and rock absorb energy in a shallow surface layer which heats up quickly and then cools down quickly. This means that continental areas have a very large temperature range – as much as 40°C/70°F in parts of the USA. By comparison there is a low range in the UK, which has a maritime climate because of its proximity to the sea. Here the temperature range is as low as 10°C/50°F.

Few parts of the Southern Hemisphere have a true continental climate but the Veldt of South Africa and the Pampas of Argentina have climates that bear some resemblance. However, they have slightly more temperate climates, being a little closer to the sea, and closer to the Equator, they have slightly warmer winters..

Below: Kalahari landscape and camelthorn in South Africa's veldt, Kalahari Gemsbok Park. The word veldt comes from the Afrikaans language and literally means 'field'. The wide open veldt are usually covered in grass and low scrub.

Above and below: Prairies are areas of land that historically support grasses and herbs with few trees. The prairies of North America, have temperate climates and moderate rainfall. The word prairie comes from the French word for meadow.

Maritime: The Effects of the Sea

Oceanicity is a measure of the degree to which a region's climate is influenced by the flow of air from over the sea. A maritime west coast climate is typical of the west coasts of mid-latitude areas but is also found in a small part of south-east Australia.

This climate type is characterized by cool and moist summers and mild winters, with a small annual temperature range. It has no serious extremes, unlike those of continental climate areas such as the Russian Steppes.

A maritime climate, such as that of the UK, is moderated by the influence of the sea. Water has a high specific heat capacity which means that it takes a long time to warm up in the summer but retains heat for some time and so takes a long time to cool in the winter. The highest land temperatures occur in the summer months of July and August. The maximum sea temperatures, however, occur in September when the air temperatures are falling. This means that areas near to the sea benefit from the air blowing across relatively warm seas in the autumn and winter so maintaining air temperatures across the land. Conversely, when continental areas swelter in very high summer temperatures, maritime areas benefit from the modifying cooling of the air blowing from the sea.

The British Isles benefit from a moderate maritime climate as the prevailing south-westerly winds blow across the Atlantic. The maritime air masses from this direction are mild due to the warming effect of the Gulf Stream, a warm water current. In Britain the average annual temperature range is only about 10°C/18°F.

Alaska and western Canada experience a maritime climate but, unlike the UK, they have no warm water current to keep temperatures up. Instead, these areas are generally rather colder and have more precipitation falling as snow.

Below: Beargrass is found in maritime west-coast climate zones such as Washington State, USA. This climatic zone generally has rainfall throughout the year, mild winters and moderately cool summers. These conditions are only found down a narrow band on the west coast of America, but cover most of western Europe.

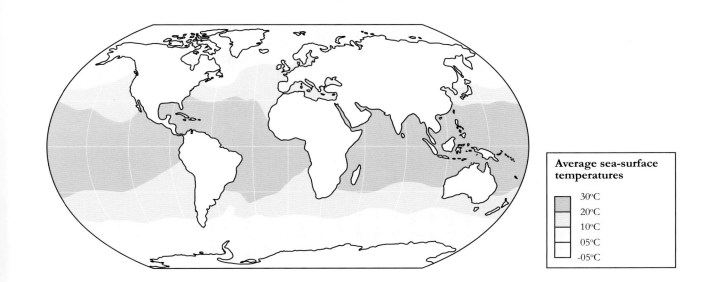

Average sea-surface temperatures

	30°C
	20°C
	10°C
	05°C
	-05°C

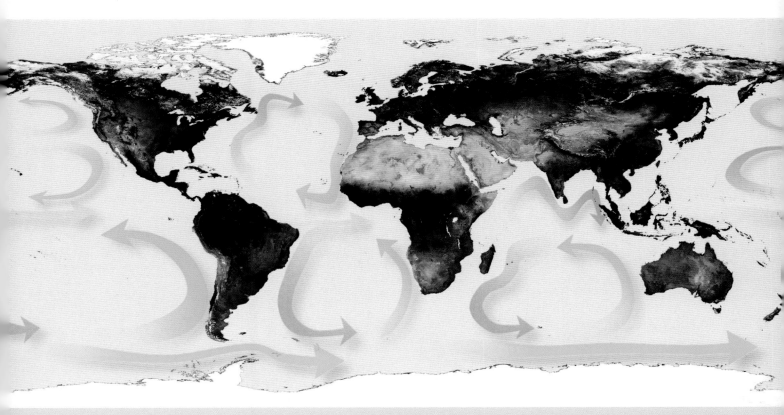

Above: Computer artwork showing surface ocean currents. Warm water is red and cold water is blue. Surface currents are driven by the winds. At bottom is the eastward flowing Antarctic Circumpolar current. It is unobstructed by land for the entire circuit of the Earth and is the strongest current. The Gulf Stream is a warm current that arises in the Caribbean Sea and travels up the Eastern coast of the USA before crossing the Atlantic. It keeps western Europe significantly warmer than it would otherwise be. The Trade Winds drive the water westward along the equator and at mid-latitudes the water is driven eastwards by the easterlies. These opposing winds cause the currents in the ocean basins to form gyres, or giant loops.

Altitude

Anyone who has flown recently may have noticed changing temperatures in the atmosphere as the aircraft climbs to cruising altitude. On the ground we can see similar decreases of temperature with altitude as we climb up a mountainside, for instance. This decrease of temperature with altitude is called the normal lapse rate of temperature or environmental lapse rate of temperature. The normal lapse rate, on a global scale, equates to 6°C in every 1000 metres/3.3°F in every 1000 feet; however, on any given day at any given location this may vary, at which point it is called the environmental lapse rate. The decrease in temperature is caused by increasing distance from the source of heated air, that is, the Earth's surface. The atmosphere is warmed by the land, which is itself heated by incoming insolation from the sun. Air is warmer nearer the Earth's surface as it is closer to the source of heat.

When air rises through the troposphere the pressure exerted on it decreases; consequently the air expands in volume. As it does so the molecules spread out and collide less with each other and cause the air's temperature to decrease. This is called adiabatic cooling. It is similar to what happens when air is let out of a tyre. The air can be felt to cool as the air rushes out and moves into a lower pressure environment; in doing so it expands against the outside environment air. The reverse is also true: as air descends through the troposphere it is put under increasing pressure which causes the air to decrease in size and squeezes the air molecules closer together. As the air volume shrinks the molecules bounce off each other with great speed and so increases the temperature of the air. This is known as adiabatic warming.

The rate at which air cools or warms depends on how much moisture is in the atmosphere. If it is dry the temperature change is 10°C for each 1000 metres/5.5°F for every 1000 feet and it is known as the dry adiabatic rate or DAR. This is always a constant, that is it is always 1°C in every 100 metres/0.5°F for every 100 feet. If the air is saturated the rate of change is 6°C for each 1000 metres/3°F for every 1000 feet and is called the saturated adiabatic rate or SAR. This varies a little depending on how much moisture is in the atmosphere. That there is a difference in the two rates is down to the release of latent heat as a consequence of condensation. As water vapour condenses it releases latent heat which is transferred to the other molecules inside the parcel of air leading to a reduction in the rate of cooling.

Orographic Effects

Those who live in hilly or mountainous areas are aware that the weather they experience is often different from that of the surrounding areas. This is what orography is, the effect of those hills and mountains on weather.

As the wind blows over the sea it picks up moisture from evaporation. If the air is then blown onshore it is forced to rise when it reaches the hills. As it does so it cools and allows condensation to take place which causes the formation of clouds from which precipitation falls. Most rain falls on the hill tops or the windward side of the hills, the side facing the sea and so the onshore winds.

As the air descends on the lee or sheltered side of the hills it warms; evaporation takes place and the air dries so little rain falls. This results in the lee sides experiencing much drier conditions than the windward sides.

This is important to understand when preparing forecasts or planning for such regions. In many parts of the world where this situation exists it is common to find the land in the lee of mountain ranges to suffer from insufficient water and to be more prone to drought than windward areas.

It is also possible for hills and mountains to produce clouds by themselves. Some mountains develop lenticular clouds which are formed by locally rising air which condenses as it rises over the top of the hill. This air cannot rise further, however, and so the winds then 'bounce' downstream of the mountains. These lenticular clouds do not move along with the wind but appear stationary. They form as the air rises and dissolve as the air descends hence the apparent stationary effect. Isolated peaks such as, famously, the Matterhorn in Switzerland develop banner clouds. As the air flows up the mountain an eddy develops in the lee of the summit and as the pressure falls it creates a lenticular cloud.

Above: Leeward sides of hills are drier than the windward.

Right: Mountain cloud forming over a coastal mountain in the Arctic summer. This cloud is formed as a result of moist sea air rising to high altitudes when it meets the mountain. The water vapour condenses in the cold temperatures at these altitudes, forming orographic clouds. The resulting rain supports life in this otherwise arid desert of the Arctic.

Opposite above: The Namche Bazaar is a village in the Khumbu region of Nepal. It is located at 3440 metres/ 11,286 feet on the side of a hill.

Opposite below: The summit of Mount McKinley in Denai National Park, Alaska, is 6194 metres/20,320 feet high. The air at high altitude is cold with a low atmospheric pressure. This has a profound effect on mountain climbers The higher you climb, the less oxygen there is in the air and if care is not taken to acclimatize to the conditions altitude sickness can occur. It commonly occurs above 2500 metres/8000 feet and can lead to the fatal high-altitude pulmonary oedema.

Local Winds

In many parts of the world the local geography, such as mountain ranges, hills and valleys, can produce a wide range of local weather conditions including winds – sometimes seasonal and with potentially devastating consequences. Many of the winds have been given names which highlight the nature of the effects they bring.

Perhaps one of the most famous is the strong, cold wind that blows south off the Alps, the Mistral. This usually develops as a cold front which blows southwards across France to the Mediterranean. Air piles up in the Alps then overflows and spills down the Rhone valley at speeds of up to 150 kph/93 mph. Marseilles and coastal resorts such as St Tropez often feel the full force of this wind when it reaches the sea. It usually brings dry, colder summer weather and high seas – a godsend for surfers along the Mediterranean coastline. An unwanted side effect, however, is that the wind can fan the summer forest fires of Provence.

In the USA the Chinook wind or 'snow eater' blows off the Rocky Mountains. This creates a warm wind as the air descends down the lee side and becomes compressed, in other words adiabatic conditions. It marks the end of the winter's snow and heralds the start of spring.

The Santa Ana wind of California blows during autumn and spring bringing hot and dry weather to the area. It is warm because of the heating of the air during its descent from the Great Basin between the Sierra Nevada and the Rockies. It commonly reaches speeds of about 65 kph/40 mph, but when it flows between mountain passes it can approach hurricane force. The hot and dry conditions can exacerbate the bush and forest fires that develop in California and can also bring dust storms from the desert to the Californian coastal cities.

In contrast, in an already windy part of the world, a cold wind blows from the Andes in a South Westerly direction across Argentina's Patagonia district. This is the Pampero wind which brings stormy conditions and a significant drop in temperature.

Some local winds are associated with changes in human behaviour and perhaps the most well known is the Fohn of the Alpine area of Europe. This is a warm, dry wind which has been associated with significant increases in reported headaches, depression and even suicide.

Left: Aerial view of the Rockies, which have a major effect on the country's weather, acting as a huge dam between eastern and western air masses. Wintertime cold air may be dammed against the Rockies for days on end, eventually to be dislodged by a burst of warmer air from above. The contrasts in temperature make for highly unstable conditions.

Below left: The Mistral is a cold dry wind blowing from the north over the north-west coast of the Mediterranean. Waves associated with the Mistral are commonly 4.5–6 metres/15–20 feet, but can reach a height of 9 metres/30 feet. The Mistral is most common in winter and spring.

Below: The shape and topography of the land and sea in many parts of the world combine to form marked local winds which have been given special names over the years. The Chinook, in the United States and the Sirocco in North Africa are some of the most famous examples of local winds.

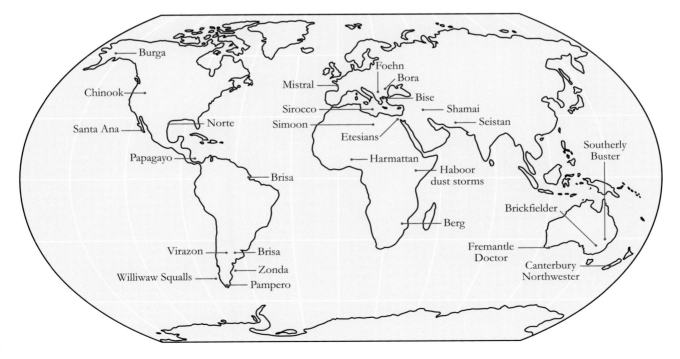

Microclimates

If you have ever noticed how a south-facing garden is warmer than one facing north then you already have some idea of what a microclimate is. 'Microclimate' refers to the state of the atmosphere in a small area of the Earth's surface, perhaps a forest, town or even a garden.

Natural microclimates occur widely, such as in an area of forest. The treetop canopy in areas like the tropical rainforests of Brazil is thick and dense. This absorbs solar radiation and so develops very high temperatures. Beneath the canopy temperatures gradually decrease because of shading; in summer the temperature difference can be up to 5°C/40°F. Humidity levels and night-time forest temperatures are usually higher than those of the surrounding area and winds are light because of the protection from the canopy.

Man-made microclimates are common. Grand Victorian houses often had walled kitchen gardens which offered protection from wind. Fruit trees and tender plants would have been grown against south-facing walls, which acted like storage heaters, absorbing solar energy during the day and re-radiating it at night, helping to maintain temperatures and assist ripening. Modern gardeners apply similar principles, growing tender species against south-facing brick walls to take advantage of their garden's microclimate.

Right: Microclimates are used to advantage by gardeners who carefully choose and position their plants in purpose-made environments to protect delicate species.

Right: A microclimate is a local external atmospheric zone where the climate differs from that of the surrounding area. Examples include the tropical rainforests. The temperature and humidity underneath this rainforest canopy will be very different from those of the surrounding external area.

Opposite: In the Eastern Highlands of Zimbabwe the waterfalls at Bridal Falls are the second highest found in Africa. Geographical features such as seas or oceans, and hills and mountains all have a significant effect on the climate. Places in mountains are normally colder than places on the plains, with more rain, snow and wind. Places in the lee of mountains are often drier because the clouds drop their rain in the mountains.

Urban Heat Islands

Perhaps the most significant man-made microclimate is the urban heat island. As towns and cities have grown so the difference in temperatures between them and the surrounding rural areas has increased. Brick and concrete buildings and dark tarmac surfaces absorb solar radiation during the day. At night this heat is released and warms the atmosphere. Heat from factories, air conditioning and refrigeration units, central heating and vehicle exhausts all add to the problem. Tall buildings also tend to trap the heat between them; roads and pavements do not retain moisture – it is channelled away by an extensive network of drains – so they remain warm. The temperature differential is obvious all year but is often more pronounced during the winter. In any case temperature differences of up to 10°C/18°F have been recorded. A characteristic of the heat island effect is that when rain falls in summer it can be in the form of torrential storms as a result of the process of convection. Water vapour requires

a 'hygroscopic nucleus' to condense around to form raindrops; this is usually a particle such as dust or grit, common in the atmosphere over cities. Combined with the very high temperatures over cities, this creates rapidly rising, moist air which condenses, forms clouds and generates thunder storms. In the USA, NASA research has shown that some cities are prone to very violent storms with thunder and lightning which the surrounding rural areas do not receive.

Above: The heat island effect can sometimes result in torrential rain which causes flooding. In this picture residents are removing flood-damaged furniture from their homes. On 8 January 2005, heavy rain fell in Carlisle in Cumbria, England. The equivalent of one month's worth of rain (over 100 millimetres/4 inches) fell in 36 hours, causing the River Eden and its tributaries to overflow severely, resulting in the worst flood for more than 100 years. Hundreds of people were evacuated as homes flooded and roads were closed. Three people died and millions of pounds worth of damage was caused to nearly 3000 properties.

Above: Thermogram showing the distribution of heat over city buildings. The colour coding ranges from yellow and red for the warmest areas (greatest heat loss) through pink to purple and green for the coolest areas (lowest heat loss). Typically roofs and windows show greatest heat loss. Thermograms are often used to check buildings for heat loss, so that they can be made more energy-efficient through improved insulation.

Right: Urban heat islands can enhance convectional uplift and the strong thermals that are generated during the summer months may serve to generate or intensify thunderstorms over or downwind of urban areas.

Temperature Inversions

The normal pattern in the atmosphere is for temperatures to decrease with height. This is because the ground is heated by solar radiation which then warms the air above it due to radiation. Under certain conditions, however, the pattern is reversed and temperatures can be colder at ground level than in the atmosphere above it, creating a temperature inversion. There are several reasons why these may form.

In California citrus fruit growers worry about clear skies on winter nights since the day's relatively warm air can radiate away and temperatures may drop below freezing. Along the California coastline wind blowing onshore may have moved across cold water causing its temperature to drop. When it reaches the coastline the cold, dense air cannot easily rise so displaces the lighter, warmer air, resulting in a temperature inversion. This has been responsible for the persistence of smog over cities such as Los Angeles, as vehicle exhausts and other sources discharge pollutants into the atmosphere. The cold, dense air of the inversion means that they are unable to disperse and are instead kept close to ground level.

In valleys at night another type of inversion can take place. The air over hills is cold and, under the influence of gravity and because cold air is dense and heavy, it flows down into the valley where it displaces the relatively warmer air. The movement of the air is a katabatic wind, which is light and can often go unnoticed. The result is sometimes known as a frost hollow, because the likelihood of frost in the valleys is very high. One of the most remarkable examples lies in the Chess valley near Rickmansworth, northwest of London. Frost has been experienced in every month of the year as cold air flows down off the Chiltern Hills into the deep, narrow valley.

Above main picture: Aerial view of the anvil-shaped top, or thunderhead, of a cumulonimbus storm cloud. The thunderhead forms at the temperature inversion layer in the atmosphere, where the air above is warmer than the air below. The rising columns of moist, warm air which create cumulonimbus clouds are trapped by the inversion layer and so spread outwards. Cumulonimbus clouds can be up to 15 kilometres/9 miles tall and appear dark from beneath. The interior of a cumulonimbus cloud is highly turbulent, the built-up energy and water being released in the form of lightning and torrential rain or hail.

Above inset: Photochemical air pollution, or smog, over Los Angeles, California, USA. This results from hydrocarbons and nitrogen oxides, mainly from vehicle exhausts, reacting together in the presence of sunlight. The pollution limits visibility, damages plants and irritates the eyes. High levels of smog can be extremely dangerous, especially to the very young and old, and those with existing respiratory complaints.

Pollution and the Weather

Left: This strange landscape is the result of different types of street lights, from several small towns and villages, diffusing through fog. The view was taken from Colli Euganei near Padua in Italy. The colour temperature of different types of public lighting causes the fog to glow in a variety of different colours. The yellow areas are associated with ordinary filament lighting, the red areas with sodium vapour lights and the blue ones with mercury vapour lights. The picture is also an illustration of the high levels of light pollution this is reached even in rural farmland areas.

Below: View across part of the city of Shanghai, China, showing very thick smog. Various emissions from many motor vehicles – particularly unburned hydrocarbons – have reacted with low-level ozone to form this brown, hazy smog.

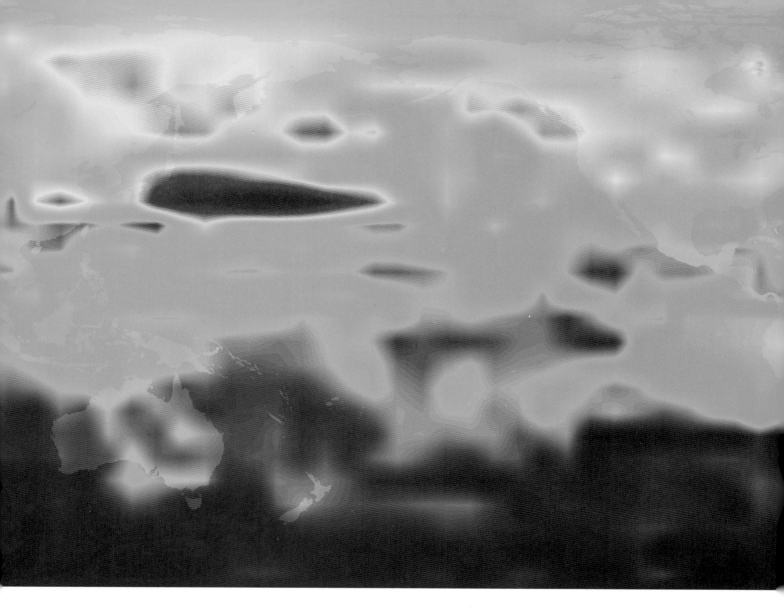

The London Smog of 5–9 December 1952

Writing in Bleak House in 1853, Charles Dickens described the characteristic smogs of the city as 'London Particulars', as they were so common. They were not new; smogs, a mixture of fog and pollution, sometimes called 'pea-soupers' because of their colour and thickness, had existed since Elizabethan times and were even referred to by the diarist John Evelyn in the 17th century. He campaigned against the burning of coal in the city because of its bad polluting effects.

It was not until the winter of 1952–3, however, that events took such a serious turn that it became essential to tackle the problem. During the immediate post-war period not only was coal still the most common fuel burned in homes and power stations in the city but diesel was being used to run the buses that were replacing trams as the main form of public transport. The early part of the winter of 1952 was particularly cold and people were burning more coal than normal to keep warm. The cold spell was due to a high pressure system which remained over the city from the start of December. London lies in a broad, shallow basin and a katabatic wind caused the very cold air from the surrounding hills to sink into the basin. The relatively warmer air over the city was forced upwards creating a temperature inversion. As the air in the inversion was calm and saturated, all the pollutants from factories, power stations, homes and transport were unable to be dispersed and mixed with the fog that developed at the same time.

Effects

Visibility fell to below 10 metres/30 feet for almost 48 hours at Heathrow airport but in places people could not see their feet! Transport came to a standstill and a performance at the Sadler's Wells Theatre had to be cancelled as it became impossible for the audience to see the stage and the performers to see each other. The emergency services and mortuaries were overwhelmed: at least 4000 people are known to have died, mainly as a result of bronchial or cardio-vascular conditions worsened by the smog. At Smithfield Cattle Market in central London it was reported that the animals were dying of asphyxiation. The death rate peaked at 900 a day on 8 December, but over 2000 died in the week ending 6 December alone. Many researchers have since put the number of deaths unofficially at over 12,000.

The government was forced to take action and the Clean Air Act of 1956 was passed making the 'pea-soupers' a thing of the past. Despite the legislation, vehicle exhaust emissions remain a principal cause of London's urban pollution problems today.

Donora, Pennsylvania, USA

In 1948 the steel- and zinc-making town of Donora suffered a serious industrial pollution incident which killed 19 people in just 24 hours. The small town of about 14,000 people lies in a river valley and that October a temperature inversion resulted in a persistent fog developing which, mixed with the industrial pollutants released into the atmosphere, caused highly noxious smog to form. Of the 19 who died, almost all suffered from chronic heart disease or bronchial problems. Another 500 became ill from similar medical conditions which worsened during the two days at the peak of the smog, with as many as 7000 people believed to have been affected in total. The problem only eased after a temporary closure of the offending factories. Although not directly responsible, the event helped to put pressure on national and state government policy which eventually resulted in the Clean Air Act of 1970.

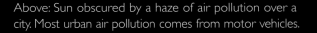

Above: Sun obscured by a haze of air pollution over a city. Most urban air pollution comes from motor vehicles.

Opposite: Satellite image of atmospheric carbon monoxide (CO) flowing east across the Pacific Ocean at an altitude of 4500 metres/1500 feet. Red indicates areas of high concentration (approximately 250 parts per billion) and blue indicates areas of low concentration (approximately 40 parts per billion). Carbon monoxide is caused by the incomplete combustion of carbon in, for instance, forest fires and petrol engines. It is an important tropospheric (lowest region of the atmosphere) pollutant. Images taken by NASA's TERRA satellite show the carbon monoxide's progression over six days.

THE EFFECTS OF WEATHER

Man's Reactions

Most of us will have heard the old saying 'it's an ill wind that blows nobody any good' but does the weather influence humans in this way? There is an increasing amount of evidence to suggest that in fact the weather does indeed have an effect on our behaviour.

Of course, we all feel better when the sun is shining, summer or winter; however, lack of sunlight is believed to be the cause of 'the winter blues' or Seasonal Affective Disorder (SAD). Sunlight stimulates the hypothalamus, that part of the brain that controls the body's main functions such as mood, activity and sleep. If the body is not exposed to sunlight these functions slow down and people feel lethargic or depressed.

It is in the more extreme latitudes that the numbers affected by SAD are highest, for instance Alaska where research suggests that 10 per cent of the population suffer, compared with Florida with an estimated 1 per cent.

Treatment is, obviously, with sunlight – phototherapy – but since the reason so many are affected is because sunlight is absent, light boxes are used instead. Sufferers use them for several hours a day in order to counteract the effects of long winter nights.

Wind

Ask any school teacher about how windy days affect their pupils and they will tell you that the children are noticeably more excitable. On a different scale, however, seasonal winds bring the reputation of being those 'ill winds' of the proverb. The Fohn and Mistral (Europe) and Chinook (USA) winds can increase temperatures by as much as 15°C/38°F in only two hours. In France it is believed that the Mistral season coincides with an increase in the suicide rate. According to a study in 2004 by the Allensbach Institute in Germany, one third of their respondents blamed the Fohn wind for adversely affecting their health. Although no one can fully explain these effects, it is thought that it may be due to the electrical charge in the atmosphere. When people are exposed to negative ions in the air they report feeling energetic and positive, and vice versa. Warm winds such as the Fohn and the Chinook are positively charged.

In developed societies people are spending increasing amounts of time indoors in centrally heated or air conditioned environments, meaning that the body is less exposed to the changing weather and so may be less able to adapt to different conditions.

Right: A woman is reading a book as she undergoes phototherapy in front of a light screen. Phototherapy is used to treat jet lag and Seasonal Affective Disorder (SAD). SAD causes depression and withdrawal in winter months, as daylight time decreases. By exposing the patient to a strong light source for a few hours each day, the impression of longer daylight hours can be created.

Wind Chill

If you have ever noticed how much colder it feels when the wind blows, as if it 'goes through' you, then you have experienced the wind chill effect. It is a measure of the amount of heat lost from the skin as the wind blows over it, in other words how much colder we feel the temperature to be as a result of the wind blowing. For instance, when the temperature is 0°C/32°F a 10 knot wind will make it feel as though it is –5°C/23°F. An increase in wind speed to 25 knots will cause the temperature to feel as though it is –12°C/10°F.

    The idea was developed to take into account the conditions likely to be experienced by those working in Antarctica. The obvious problem there is frostbite and research showed that wind speed was critical to the development of the condition. Scientists realized that it was quite possible to work in temperatures as low as –40°C but that it would only take a wind speed of 3 or 4 knots to make an enormous difference.

    Perhaps one of the most extreme examples of the wind chill effect occurred in the USA in January 1977 when the city of Cincinnatti recorded –25°C/–13°F, the coldest since records began in 1820. On the morning of 28 January, however, Minneapolis recorded a wind chill temperature of –61°C/–78°F.

Right above: Scientist measuring the wind speed in Antarctica using a hand-held anemometer. The anemometer has several small cups which are driven round by the wind. The faster the cups move, the greater the output from the anemometer.

Right below: Wind chill is the apparent temperature felt on exposed skin due to a combination of air temperature and wind speed. Except at higher temperatures, the wind chill temperature is always lower than the air temperature, because any wind increases the rate at which moisture evaporates from the skin and carries heat away from the body.

Plant and animal adaptations

Plants and animals demonstrate a wide range of physical adaptations and behavioural strategies for surviving and thriving in particular environments, habitats, and biomes – which consist of large communities of plant and animal species, in which the latitude and climate are major determining factors.

In temperate biomes or climates, for example, conditions may be highly changeable, particularly between the seasons of summer and winter. In polar and tropical climates, conditions may be more stable or more extreme, especially with regard to temperature, humidity and rainfall – in tropical and subtropical regions rainfall can vary seasonally, with extreme differences between the wet and dry seasons.

Additionally, coastal areas, valleys, mountains, forests and cities may also experience microclimates, where climate and weather conditions may be affected by the lie of the land, the amount of vegetation present, large bodies of water or conurbations, which may affect sunlight, ground and air temperatures and wind patterns.

Plants typically display adaptations in their leaves and roots, the forms of which may enable them to best obtain sunlight, retain moisture, survive extremes of temperature and humidity, protect them from severe winds, and allow them to thrive where nutrients are in short supply.

In animals, the main adaptation in relation to climate is probably that of blood temperature. In ectothermic, or 'cold-blooded' animals this is governed by the temperature of their surroundings. In endothermic, or 'warm-blooded' animals, it may be determined to an extent by the ambient temperature, but relies primarily on internal temperature regulation, which is achieved by processing carbohydrates and body fat. Ectotherms, such as reptiles, may therefore bask in the sun to warm themselves, whilst endotherms must eat to stay warm in cold conditions, and may possess thick layers of insulating fat, dense feathers or fur, and pant or sweat when too hot. Animals may also seek refuge underground when too hot or too cold, or enter shade or water to cool themselves. Some may hibernate for extended periods when it is extremely cold, slowing their metabolisms and lowering their body temperatures, whilst others, particularly birds, may migrate to warmer climes.

Opposite: Many species of tree adapt to the conditions of the climate of the places where they grow. The leaf stalks of many trees that grow in hot climates, such as the Australian and Pacific islands and Africa, are vertically flattened; this helps to protect them from intense sunlight. In temperate climates decidous trees respond to winter by dropping their leaves.

Right above: Only animals that have adapted to the long cold winters and harsh conditions in the Antarctic can survive. Seals have a layer of blubber – a thick layer of fat between the muscle and skin that varies with age, body size and energy balance. It is built up when food is abundant and used as an energy source when required.

Right below: A marine iguana is an ectotherm – a 'cold-blooded' creature. It spends much of its time feeding on seaweed underwater. Due to the coldness of the waters it inhabits, it basks in the sun to raise its body temperature. The marine iguana is only found along the rocky coastline of the Galapagos Islands.

Temperate

In the temperate regions between the poles and the Equator, four seasons are experienced, presenting a variety of conditions that plants and animals must cope with. These include warm, humid summers, and cold, often snowy winters, particularly in northern temperate zones, where freezing conditions may persist for much of the year. Rainfall tends to average about 610–1520 millimetres/24–60 inches.

Temperate regions tend to be dominated by deciduous broadleaf forests, containing trees such as oak, maple, birch and beech, coniferous forests, consisting of cedar, spruce, cypress, firs and other evergreen varieties, and mixed forests, which contain both types of tree.

The needle-like leaves of conifers help to minimize water loss, and are protected against freezing by large spaces between their cells, whilst the leaves of deciduous trees tend to be large and flat to maximize their exposure to sunlight, and are shed in the winter to prevent excessive moisture loss, at a time when water in the soil may be frozen.

Birds and mammals of temperate areas may migrate seasonally in winter in order to find more favourable conditions, whilst some mammals, reptiles and amphibians, may hibernate. Other mammals may remain active in winter, sometimes relying on stored food, and some may develop a dense winter coat, which in some species in northern latitudes, such as the snowshoe hare, is white, in order to camouflage them against snow. Many insects, however typically die in the winter, having laid eggs during the summer and autumn, which will hatch the following spring or summer.

Above: The grass snake hibernates in winter in order to conserve energy and help survive winter conditions when food is scarce. The grass snake is the longest reptile in Britain reaching up to 20 metres/65 feet in length. Its main form of defence is to fake death when it feels threatened – another effective conservation strategy.

Polar

The polar regions are characterized by extremely low temperatures, low rainfall, strong winds, and snow and ice. During the winter, there are long periods with little or no sunlight, whilst in the summer there are extended periods of daylight, although the sun remains low on the horizon, and temperatures only rise above freezing for a few months, perhaps reaching around 12°C/54°F at best.

The dominant habitat is tundra: essentially frozen marshland or bog, with low-growing grasses, sedges, lichens and mosses being the main forms of vegetation. The nutrient-poor topsoil is constantly freezing and thawing, whilst the subsoil is permafrost, which prevents the establishment of deep root systems and prevents soil drainage. Plants grow close to the ground where it is slightly warmer, and to escape wind damage, and may possess hairy leaves to trap warmth. Root systems are shallow, but may branch horizontally to hold plants in place and seek nutrients, whilst lichens are able to obtain their nutrients from the air, and may grow directly on bare rocks. They are also able to photosynthesize at very low temperatures.

Terrestrial fauna is limited in polar habitats, and is particularly restricted in the Antarctic, although there are several species of marine mammals that inhabit both polar regions, and which are protected against the cold by dense layers of fat or blubber. Similarly, terrestrial Arctic mammals, such as the musk ox, polar bear and arctic fox, possess insulating fat and very dense fur to keep them warm. Many birds also breed on the Arctic tundra in summer, feeding on the abundant insect life, but most migrate south in winter. Those that do not, such as the ptarmigan and snowy owl, possess dense, white plumage. The Emperor penguins of the Antarctic are able to withstand the coldest temperatures on Earth by huddling together in huge colonies, protected by dense layers of both feathers and blubber.

Below: Penguins are particularly well adapted for a life in the cold Antarctic waters. Heavy, solid bones act like a diver's weight belt, allowing them to stay underwater. Their wings, shaped like flippers, help them swim underwater at speeds up to 24 kph/15 mph. A streamlined body, paddle-like feet, insulating blubber, and watertight feathers all add to their efficiency and comfort when submerged. In addition to their insulating blubber, penguins have stiff, tightly packed feathers that overlap to provide waterproofing and they coat their feathers with oil from a gland near the tail to increase impermeability.

Arid and Semi-arid

The arid and semi-arid regions of the world are defined by dry climates, where there is typically very little rainfall, although in semi-arid regions, where the main habitats are grasslands, such as scrub, or savannah with sparse populations of trees, there may be wet and dry seasons. In truly arid habitats, however, the main biome type is that of desert, which may experience no rainfall some years. Humidity is also very low, ensuring that temperatures can vary massively on a daily basis, with the full force of the sun beating down during the day. This sometimes produces temperatures of over 50°C/122°F, and a lack of cloud cover at night meaning that temperatures can fall to –20°C/–4°F.

In the large deserts of the world, which occur mainly around the Tropic of Cancer and the Tropic of Capricorn, plant cover is very sparse, but those plants that are present tend to be succulents, with fleshy, waxy stems and leaves, designed to retain water, and extensive root systems in order to obtain it. Similarly, grasses in semi-arid areas have large root networks which seek out available moisture, and also protect against overgrazing, sending up new shoots as they spread. Large grazing animals meanwhile tend to migrate to limit overgrazing, and feed mainly at dawn and dusk, when plants retain more moisture.

In both deserts and dry grasslands, smaller animals may spend much of their time in burrows, avoiding the extremes of temperature during both the day and night. Reptiles and invertebrates are able to retain moisture on account of their scaly skin and waxy exoskeletons respectively, whilst many small mammals, such as kangaroo rats and gerbils, may obtain all their water from the vegetation they consume. Camels are superbly adapted to dry conditions, being able to store fat in their humps, and able to go for long periods without drinking water.

Above: Camels are well-adapted to life in the desert. Their hump contains fat that can provide both their nutritional and water needs over prolonged periods. Their feet are wide, allowing them to walk on soft sand, and have thick soles to withstand the ground's heat. They are able to close their nostrils against airborne dust and their eyes are protected by thick eyelashes.

Opposite above: Many mosses, lichens and ferns belong to a group of plants called epiphytes. These are plants which grow above the ground surface, using other plants or objects for support. These plants have adapted to life in climates such as a tropical rainforest, where they can reach positions where the light is better or where they can avoid competition for light.

Tropical and Subtropical

Tropical climates and habitats occur in the equatorial regions, bordered by subtropical regions to the north and south. Tropical regions experience high levels of sunlight, high temperatures and rainfall, sometimes of several centimetres/inches per month, promoting the development of luxuriant rainforests, with incredibly high species diversity. Conditions in subtropical regions may be similar, but tend to experience more seasonal rainfall, leading to the establishment of monsoon forests. Here plant growth may be limited by a shorter growing season, trees may shed their leaves during the dry season and animals may migrate to be close to sources of water.

In rainforests, where precipitation may exceed 255 centimetres/100 inches annually, causing nutrients to be leached from the soil, distinct layers of vegetation occur, from the forest floor, where trees often have spreading buttress root systems to maximize nutrient supply, and where lack of light prevents the establishment of dense undergrowth, to the tree canopy, where vegetation is more dense, with a variety of epiphytes, such as climbers, which grow on the trees, obtaining their nutrients and moisture from the air. The leaves of trees in the understory tend to be broader than in the canopy in order to utilize maximum sunlight, and many have 'drip tips', which encourage water to run off from their surfaces.

The warm, humid conditions are ideal for many invertebrates, reptiles and amphibians, and some frogs for example are able to lay their eggs in pockets of water held by plant leaves, or even to carry tadpoles on their backs. The dense vegetation also provides food and refuge for numerous birds and mammals, many of which feed on fruits, helping to disseminate seeds, while several birds and insects also pollinate flowers. Mammals may sweat and pant, and retreat to water or burrows in order to cool themselves and prevent overheating.

Right: Some frogs found in Central and South America have adapted to the absence of pools of standing water by not having a swimming tadpole stage in their life cycles. Instead, the eggs are laid on land and hatch out as fully developed froglets.

WEATHER EXTREMES

STORMS

Storms are possibly the most dramatic of the common weather events. They can bring heavy rain, sometimes hail, thunder, lightning and occasionally tornadoes or waterspouts, almost any of which can cause significant damage, particularly in urban areas. As many as 2000 storms can be present in the Earth's atmosphere at any time, mainly in tropical and sub-tropical areas during spring and summer, although not in the Antarctic. Storms are characterized by huge anvil-headed cumulonimbus clouds which can be between 2–10 kilometres/1–6 miles across and may reach a height of 19 kilometres/12 miles. Inside these clouds vast amounts of energy are generated and released. Storms usually last one to two hours but the more severe ones can last much longer.

The starting point for any storm is differentially heated air. A typical summer thunderstorm occurs because the ground surface is heated locally to high temperatures, much higher than the surrounding air, as may happen after a period of very hot weather. The warm surface heats a 'cell' of air which rises rapidly upwards (updraughts) to the troposphere creating turbulence, or unstable conditions, in other words the local air is much warmer than the air surrounding it. A large rising cell or bubble of air is formed, and clouds form as water vapour cools and condenses, forming ice crystals and water droplets. The clouds become electrically charged, both positively and negatively. As the air rises so cooler surrounding air descends (downdraughts) only to be heated again, so prolonging the life of the storm.

The storm will continue as long as the cell of air remains warmer than the surrounding air. Gradually, as the moisture in the cell is used up the supply of energy decreases, more downdraughts of air develop than updraughts and the cell dissipates; and the storm is over. Storms can also form where a cold front meets warm, moist air along a squall line, so called because the downdraughts cause gusty winds to form at the ground surface.

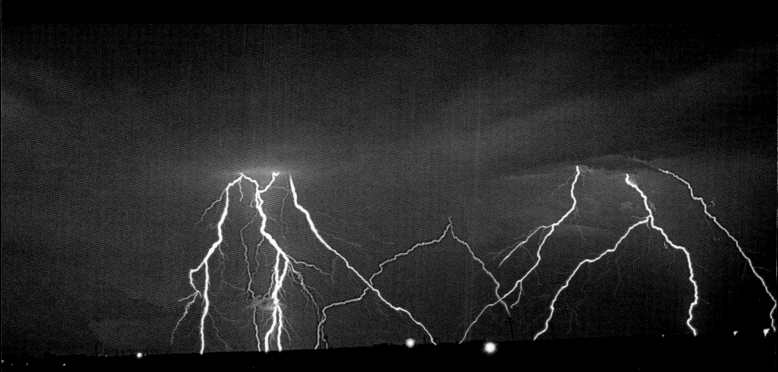

Left: Forked cloud-to-ground lightning. Lightning occurs when a large electrical charge builds up in a cloud, probably caused by the rapid movement of water droplets and ice particles in the turbulent interior. White lightning indicates that the air is dry.

Opposite below: Multiple lightning bolts striking the ground from a storm cloud at night. At first a leader of electrons descends to the ground. As soon as this makes contact an enormous return stroke surges up the leader's ionised path, producing blinding white light.

Below: Thunderstorms develop as warm, moist air rises into cooler air, causing condensation and cloud formation. As the cloud becomes increasingly unstable, the updraught strengthens. When the cloud reaches the tropopause layer of the atmosphere it begins to spread out at the top, forming an anvil shape. Mature thunderstorms can produce as much as 100 millimetres/14 inches of rain per hour, and may cause localized flooding.

Above: Tornadoes develop from intense thunderstorms that begin to spin because high-level winds are blowing faster and in a different direction to low-level winds. The rotation can be enhanced by the interaction of cold and warm air currents, causing a mesocyclone, which is a powerful air column moving upwards at the centre of the storm. From this stage, the spin will intensify and a rotating column of air will break through and hit the ground.

Below: Thunderstorms occur in warm, humid conditions when dark cumulonimbus clouds build up in the sky.

Above: A supercell thunderstorm is a severe long-lived storm within which the wind speed and direction changes with height. This produces a strong rotating updraught of warm air, known as a mesocyclone, and a separate downdraught of cold air. Tornadoes may form in the mesocyclone, in which case the storm is classified as a tornadic supercell thunderstorm. The storms also produce torrential rain and hail.

Below: Severe thunderstorm at sunset. Thunder is the sound of the shockwave caused when lightning instantly heats the air around it to up to 30,000°C/54,000°F. That super-heated air expands rapidly, then contracts as it cools. The rapid expansion/contraction generates sound waves, making the sound that is called 'thunder'.

Thunder and Lightning

Thunder and lightning are probably the most obvious and dramatic features of common storms. Although there is still no clear agreement on how lightning is formed it is believed that it's largely due to collisions between ice crystals. A huge electrical charge builds up as cloud particles become increasingly electrified and the charged particles separate. Positively charged particles develop at the top of the cloud whilst the base is negatively charged. The ground beneath the storm develops a positive 'shadow'.

The electrical charge is discharged in a flash of lightning. A weak spark, the stepped leader, falls from the cloud and a positive leader rises from the ground. The return stroke, the main discharge from the ground to the cloud, moves along a channel created by the two leaders. Air around this channel is heated to about 30,000°C/54,000°F and the electrical discharge is about 1.5 million volts, mostly converted into heat energy – a bolt of lightning is hotter than the surface of the sun. Typically a flash of lightning lasts only 0.2 seconds but generates enough heat to cause an isolated tree to explode as the water inside it is heated.

Opposite main picture: Dramatic storm clouds tinged pink with the light from the setting sun. Positively charged particles develop at the top of the cloud whilst the base is negatively charged. The ground beneath the storm develops a positive 'shadow'.

Inset above: The majority of thunderstorms occur in spring and summer in tropical and subtropical areas; they are only absent in Antarctica.

Inset centre: A shelf cloud forms when rain-cooled air from a thunderstorm drops beneath the cloud and moves out from under it. This pushes warm air up over the top of the cloud, forming a leading edge that can run the whole length of a storm system.

Inset below: The intense heat from a flash of lightning causes the air to expand and then contract violently, producing thunder. The number of seconds between flash and thunder divided by three gives the approximate distance away of the lightning in kilometres (divided by five gives the approximate distance in miles).

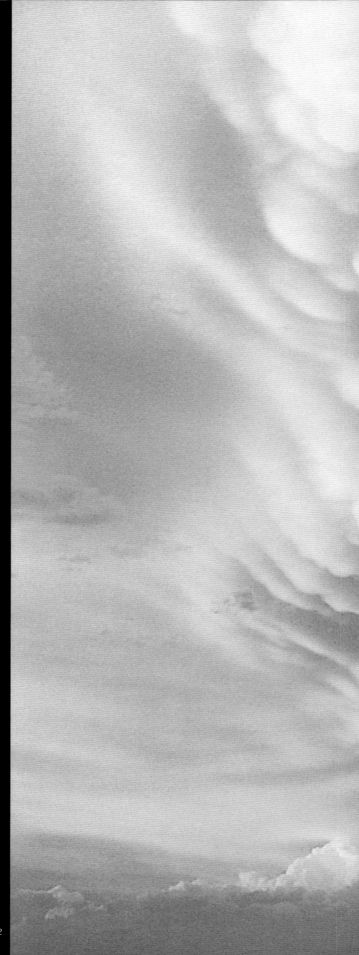

Heated air expanding very rapidly creates sound waves – thunder. The speed of sound is much less than that of light so there are several seconds' delay between the flash of lightning and the sound of thunder. This is useful if we want to work out how far away a storm is. The difference in time between a lightning flash and thunder sounding is three seconds for each kilometre or five seconds for each mile.

Lightning tends to strike at tall objects, such as isolated trees or skyscrapers and, of course, lightning conductors. People who shelter beneath trees during a storm are often at severe risk of a lightning strike. Golfers are particularly prone and professional tournaments have a detailed weather forecast prepared for the duration as the consequences could be fatal. Golfer Lee Trevino was hit by lightning during a tournament in Chicago in June 1975 and was left with permanent damage to his lower back.

In a typical thunderstorm about 1000 lightning strikes take place each hour but a particularly severe thunderstorm in Quebec City, Canada, in July 1999, received the equivalent of 3000 an hour.

It is not true that lightning never strikes twice: the Empire State Building in New York is hit about 100 times each year and was once famously hit 15 times in 15 minutes.

Above: Lightning occurs when a large electrical charge builds up in a cloud, probably because of the rapid movement of water droplets and ice particles in its turbulent interior. The charge is attracted to the oppositely charged ground, and a few leader electrons travel from one to the other. When one makes contact, there is a huge backflow of energy along the path of the electron.

Right: Forked cloud-to-ground lightning over an urban area at night. Cloud-to-ground lightning occurs between a negatively charged cloud base and the positively charged ground. Rain makes the lightning appear red in colour.

Types of Lightning

There are at least six different forms of lightning, each with its own individual characteristics and colours. Fork and sheet lightning are the most frequently seen.

Sheet lightning (or intracloud lightning) can be seen when the lightning is close to the horizon and is in fact fork lightning within a cloud. The individual strikes can't be seen, but simply light up the distant cloud. Sheet lightning is the most common type of lightning.

Fork lightning can be seen moving from cloud to ground, ground to cloud or cloud to cloud. Each fork of lightning is actually two strikes travelling quickly up and down the same path. The first bolt (the 'leader stroke') zig-zags downwards and completes the circuit between cloud and ground. This is followed a split second later by a huge surge (the 'return stroke') that shoots back up the path just created. This is also the most destructive type of lightning as it often damages buildings and can kill.

Heat lightning is like sheet lightning, but is so far away that you cannot hear the thunder. The term 'heat' has little to do with temperature. Since heat lightning is most likely to be seen in association with thunderstorms on warm summer nights, the term 'heat' may have been used because these flashes are often seen when surface temperatures are high. Sometimes heat lightning can appear to be orange. This is because of particles in the air refracting the light.

St Elmo's Fire is a green or bluish haze that hangs in the air above tall objects on the ground. It is caused by lots of sparks and the electrical field that is generated by them. The name St Elmo's fire came about because sailors first saw this type of lightning when they were on watch on the masts of ships, and St Elmo is the patron saint of sailors. St Elmo's Fire has to be attached to something like a ship's mast or the tip of the wing of an aeroplane.

Ball lightning is the rarest form; very few people have seen or photographed it and scientists do not know how it forms. Unlike St Elmo's Fire, ball lightning can float around freely.

Red Sprites, Green Elves, and Blue Jets (sometimes called high altitude lightning) are upper atmospheric optical phenomena associated with thunderstorms and have only recently been documented. This type of lightning shoots out from the top of a storm cloud and usually happens just as a lightning bolt discharges from below.

Tropical Storms and Hurricanes

The devastating Hurricane Katrina which affected the Caribbean and southern states of North America in 2005 reminded us all of the devastating power of hurricanes. These are storms that are formed between the Tropics of Cancer and Capricorn in the Atlantic and Pacific Oceans as well as the Indian Ocean. They have different names depending on where they are formed. In the Atlantic they are called hurricanes, in the north-west Pacific, typhoons; in the Indian Ocean they are known as tropical cyclones while north of Australia they are sometimes called Willy Willies!

To form, hurricanes need sea temperatures of 27°C/81°F or above, and moisture and light winds. The hurricane season lasts from 1 June to 30 November in the Atlantic and from 15 May to 30 November in the Pacific. Around 80 tropical storms form every year with most – around 35 – in the south or south-east of Asia. Twenty-five affect the Caribbean and USA, with the remainder across the southern Indian and Pacific oceans. In a normal season nine or ten will develop into hurricanes in each area; four or five of these may develop into severe hurricanes. They are enormous features, often 200 kilometres/120 miles and sometimes as much as 700 kilometres/430 miles in diameter. They may last from a few days to a week or more and their tracks are notoriously unpredictable.

Characteristic Features

From space, hurricanes look stunningly beautiful but the images mask a sequence of terrifying events. Although the rate at which the hurricane moves across the earth is relatively slow, perhaps only 40 kph/25 mph, the incredible speeds within the hurricane system are perhaps its key feature. A tropical storm becomes a hurricane when it develops sustained wind speeds of 118 kph/74 mph or more but a major hurricane can cause wind speeds in excess of 248 kph/155 mph.

Right: Palm trees are being blown by violent winds.
A category 3 hurricane indicates winds between 178 kph/110 mph and 209 kph/130 mph.

Above: Strong winds cause a palm tree to bend. The terms 'hurricane' and 'typhoon' are regionally specific names for a strong 'tropical cyclone'.

The thick blanket of cloud which characterizes the hurricane from space contains unrivalled quantities of water vapour and, inevitably, exceptionally heavy rainfall.

At the centre of the hurricane is the eye, an area of cloudless, calm weather which can be as much as 50 kilometres/30 miles across. Here the air is descending and warming, meaning high pressure, calm weather and no clouds, a short respite before the remaining part of the hurricane continues on its course.

As the strong winds approach land the final, deadly component begins to take shape, the storm surge. Low pressure in a hurricane has the effect of lifting the surface of the sea and the strong winds push the water ahead of it in huge waves. When they reach the coastline, the powerful waves crash ashore in surges. These can be up to 6 metres/18 feet above normal sea level and may travel inland for many hundreds of metres where the coastline is low-lying, causing wide-spread flooding.

It is not uncommon to find tornadoes forming as the hurricane advances, adding to the hazardous conditions.

How are Hurricanes Measured?

Hurricanes are measured according to intensity, based on wind speed and storm surge severity, on the Saffir-Simpson scale. This ranges from 1 to 5, with 5 being the most severe and, if it makes landfall, with the most devastating consequences. At 5, wind speeds will be above 248 kilometres/155 miles per hour and its storm surge will be over 6 metres/18 feet above normal sea level.

| Type | Maximum Sustained Winds | | Barometric Pressure | | Storm Surge | | Damage Level |
	KNOTS	MPH	millibars	inches of mercury	Feet	Metres	
Depression	Less than 34	Less than 39	–	–	–	–	–
Tropical Storm	35–63	39–73	–	–	–	–	–
Category 1	64–82	74–95	Greater than 980	Greater than 28.94	3–5	1.0–1.7	minimal
Category 2	83–95	96–110	979–965	28.50–28.91	6–8	1.8–2.6	moderate
Category 3	96–113	111–130	964–945	27.91–28.47	9–12	2.7–3.8	extensive
Category 4	114–135	131–155	944–920	27.18–27.88	13–18	3.9–5.6	extreme
Category 5	136+	156+	Less than 920	Less than 27.17	19+	5.7+	catastrophic

Hurricane Naming

For hundreds of years, hurricanes that occurred in the West Indies were named after the Catholic saints' days on which they took place, such as 'Hurricane Santa Anna', which occurred on 26 July 1825, and 'San Felipe' and 'San Felipe the second', which occurred on 13 September 1876 and 1928 respectively, all of which struck Puerto Rico. During the late 19th and early 20th centuries, however, the Australian meteorologist Clement Wragge began to use the names of his friends, despised politicians, and historical figures and mythological characters.

Later, officials would simply refer to such storms by the coordinates of their latitudinal and longitudinal positions, but this method proved to be both slow and open to error, and so during World War II military meteorologists reverted to a system of naming and, using the convention of applying 'she' to inanimate objects such as vehicles, hurricanes were given female names. Then, in an attempt to eliminate any possible confusion, the US army and navy adopted a system of applying names from their joint phonetic alphabet, with the first hurricane being named 'Able' in 1950. However, with the introduction of an international phonetic alphabet in 1953, that system was abandoned and female names were reintroduced by the US National Hurricane Center. A list of names was drawn up, which would be used on a rotational basis each season, with the names of particularly significant storms being retired.

In 1979, following claims of sexism, male names were introduced by the World Meteorological Organization, which continues to oversee the practice of hurricane naming. That year would also herald the practice of drawing up a list of names in advance of the hurricane season, and today an alphabetical list of 21 names is used (excluding the letters q, u, x, y and z), alternating between male and female, and including English, French and Spanish names, as these are the main languages used by the people who are most commonly affected by hurricanes. If all of the common names were to be used within a single season, the letters of the Greek alphabet would then be adopted.

Name	Season
Agnes	1972
Alicia	1983
Allen	1980
Allison	2001
Andrew	1992
Anita	1977
Audrey	1957
Betsy	1965
Beulah	1967
Bob	1991
Camille	1969
Carla	1961
Carmen	1974
Carol	1954
Celia	1970
Cesar	1996
Charley	2004
Cleo	1964
Connie	1955
David	1979
Dennis	2005
Diana	1990
Diane	1955
Donna	1960
Dora	1964
Edna	1954
Elena	1985
Eloise	1975
Fabian	2003
Fifi	1974
Flora	1963
Floyd	1999
Fran	1996
Frances	2004

Name	Season
Frederic	1979
Georges	1998
Gilbert	1988
Gloria	1985
Gracie	1959
Hattie	1961
Hazel	1954
Hilda	1964
Hortense	1996
Hugo	1989
Inez	1966
Ione	1955
Iris	2001
Isabel	2003
Isidore	2002
Ivan	2004
Janet	1955
Jeanne	2004
Joan	1988
Juan	2003
Katrina	2005
Keith	2000
Klaus	1990
Lenny	1999
Lili	2002
Luis	1995
Marilyn	1995
Michelle	2001
Mitch	1998
Opal	1995
Rita	2005
Roxanne	1995
Stan	2005
Wilma	2005

Above: A list of retired hurricane names.

Opposite below: The Saffir-Simpson scale measures the intensity of hurricanes.

The Effects of Hurricanes

A hurricane is one of the most destructive forces of nature the earth experiences. Winds can gust up to 300 kilometres/190 miles per hour and storm surges can reach as high as 6 metres/18 feet, which, along with heavy rainfall, causes serious flooding. Once they reach land the energy source of the hurricane, warm seas, disappear, and it is downgraded to a tropical storm.

Tracking and Warning of Hurricanes

Specialist national hurricane warning centres exist in many parts of the world to watch the oceans for any signs of tropical storm formation. In the United States the Atlantic is watched carefully by the National Oceanic and Atmospheric Administration (NOAA). All available technology is used – satellites, hurricane surveillance aircraft, radar, buoys, weather ships – to make sure nothing is missed: many lives and livelihoods depend on getting it right. Once the information is received, forecasts are made and the results are disseminated swiftly via a wide variety of media to all vulnerable communities and to ships at sea. Radio and television are used extensively but the internet is rapidly catching up as the preferred source of information. Telephone and SMS services are also widely used, while ships at sea use fax and radio messages broadcast via satellites. For coastal communities in some countries sirens are used to warn of the impending landfall of a hurricane, prompting evacuation and other precautionary measures to be taken. In the United States the National Hurricane Center's website provides comprehensive advice and information for all in potentially vulnerable areas, from advance preparations such as securing homes to evacuation procedures.

Above left: A boy descends from a coconut tree as strong winds blow along a baywalk area in Manila.

Left: Eye of Hurricane Floyd as seen from the cockpit of a plane flown by the US Air Force Hurricane Hunters team. The Hercules plane is equipped to measure meteorological parameters, such as wind speed, humidity and atmospheric pressure. The data is then relayed to base, where computer models help predict the storm's future direction. These models have an accuracy of 70 per cent, which allows authorities to implement damage limitation procedures such as evacuations.

Above: Flood waters in a market in Manila, Philippines, after a tropical storm.

Right: Coloured three-dimensional computer image, based on satellite data, of Hurricane Mitch (white cloud swirl, upper right). Mountainous areas of Central America (green) are brown. The hurricane is off the coast of Honduras (centre, mostly obscured). North is at top. The hurricane produced winds of over 320 kph/200 mph and heavy rain that caused flooding and land-slides. Over 11,000 people were killed and damage was estimated at 5 billion dollars.

Hurricane Mitch

In October 1998 Hurricane Mitch, which began life in the southern Caribbean Sea as a tropical storm, developed into a category 5 hurricane. It first hit the northern coast of Honduras, then moved further west to affect Nicaragua, Guatemala and El Salvador. Wind speeds were as high as 300 kilometres/186 miles per hour but it was the heavy rainfall that did most damage; 1000 millimetres/39 inches of rain fell in five days. The hurricane moved very slowly through these countries causing widespread flooding, 19,000 deaths and the destruction of settlements leading to 2.7 million people being left homeless. Fifty bridges in Nicaragua alone were destroyed: for a while Managua, the capital city, was cut off from the rest of the country. The economic effects were catastrophic; coffee and banana plantations were badly damaged and even where there was anything harvested nothing could be moved due to the disruption in the transport infrastructure. An estimated $1.5 billion dollars' worth of damage occurred, severely hindering the development of these countries.

Hurricane Katrina

In August 2005 Hurricane Katrina, developing in the Atlantic, came to the attention of NOAA. It quickly became apparent that this was no ordinary hurricane but was developing into a category 5 storm and that it was heading for land. The southern USA states of Alabama, Mississippi and Louisiana were put on high alert. The city of New Orleans was believed to be most at risk because of its unusual geographical location; on the Mississippi delta and protected by levees, much of the city lay below sea level. The worst fears of the authorities were quickly realized as, even though the hurricane was downgraded to category 3 before it hit land, the coastal communities were devastated by winds, storm surge and flooding. Eighty per cent of the city of New Orleans was flooded as the levees were breached by the storm surge; millions were evacuated to neighbouring, safer communities but many could not or would not leave and over 1800 people died. More than $81 billion of damage was estimated to have occurred. The Gulf oil industry was temporarily closed because of the wind damage and contributed to the rising price of crude oil on world markets: this hurricane had far-reaching effects.

Opposite above: On 25 August 2005, pedestrians battle against hurricane Katrina as it brings heavy storms to Fort Lauderdale, Florida, USA. Although a voluntary evacuation order was in place in Florida, many people remained and two were killed by falling trees. Katrina did more damage further west, however, as the strongest part of the storm made landfall on the Gulf of Mexico coast. It devastated coastal regions of Mississippi and Louisiana. The poor and low-lying port city of New Orleans was left under polluted flood waters, and hundreds of people died.

Opposite below: Workers wearing masks and overalls to protect themselves from chemicals clear up clear up some of the debris created by Katrina.

Below: Even concrete structures suffered serious damage when Hurricane Katrina struck. Katrina is thought to have been the most costly natural disaster in US history.

Tornadoes

Anyone who has ever read or watched *The Wizard of Oz* will have some idea of the destructive power of a tornado. The story is set in Kansas, one of the USA states which form the so-called 'Tornado Alley', an area renowned for the large number of highly damaging and life-threatening tornadoes which occur there every year. No wonder, then, that understanding when and where they may form is so important.

A tornado is a rapidly spiralling column of air, usually associated with thunderstorms. They are generally small in scale, between a few metres/feet to hundreds of metres/feet across, and can last for only a few minutes or up to an hour, perhaps longer.

The characteristic feature of a tornado is the very powerful winds which spiral rapidly at speeds that can exceed 200 kmp/125 mph. Inside the column of air a powerful updraught causes very low pressure on the ground surface while outside the column the winds rotate, anti-clockwise in the Northern Hemisphere, clockwise in the Southern.

Although the mechanisms responsible for their formation are by no means fully understood, tornadoes are likely to form when cold, dry air comes into contact with warm, moist air: this situation could occur along a cold front. Powerful thunderstorms can form along the front and can form very strong updraughts. With height, the wind changes direction and the updraft of air begins to rotate, a process known as wind shear. The rotation extends downwards and creates the recognizable funnel-like shape. If conditions are favourable the funnel reaches the ground and becomes a tornado. Very low pressure inside the column of air causes moisture to condense and makes it visible.

In the USA tornadoes are measured on the Fujita or F-scale, which goes from 0 to 5, whereas in the UK they are measured on the Torro or T-scale which goes from 0 to 10. With both scales it is wind speed that is taken into consideration when calculating which level of the scale a tornado reaches.

The most damaging tornado to affect the USA occurred in March 1925 in Tornado Alley. The towns of Murphysboro, Gorham and DeSoto were affected by the F5 tornado which caused 695 deaths and 2027 injuries. Events like this mean that a huge amount of effort has gone into research into the meteorological conditions that cause tornadoes. Warning systems are in place via a wide range of media including, increasingly, the internet. Advice is available for organizations such as schools and hospitals, as well as for private individuals, on safety and construction of storm cellars.

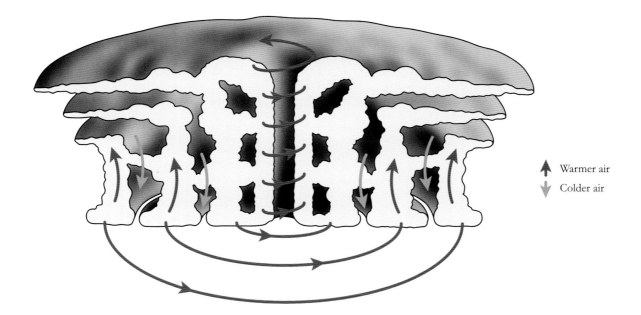

↑ Warmer air

↓ Colder air

Right: Tornadoes are far more destructive than landspouts, waterspouts or funnel clouds, but are also much rarer.

Below: In the most severe tornadoes, winds travel at over 300 kph/185 mph, causing devastating damage.

Opposite: A tornado is a rapidly rotating funnel of air that can form below certain types of storm clouds, and can be accompanied by lightning and rain. If it touches the ground, the high wind speeds in the rotating funnel can cause immense destruction, throwing up a cloud of debris. Large and powerful tornados can have wind speeds of over 200 kilometres/125 miles per hour, be many metres/feet wide at their base, destroy houses and trees, and kill people in their path. Tornadoes often form as weather fronts collide over large plains, such as the Great Plains of the USA.

Tornado Alley

The part of the USA which is most prone to damaging tornadoes is the mid-western states between the Rockies and the Appalachians, although tornadoes have been reported in every state. 'Tornado Alley' is particularly vulnerable because it lies between the hot and humid air of the Gulf of Mexico and the cold and dry air from the Rockies. This leads to the formation of enormous cumulonimbus clouds, as much as 18,000 metres/60,000 feet high. Sometimes these group together and form 'supercells'. The most dangerous tornadoes form from these, along with thunderstorms and hail showers. In April 1974 just such conditions developed to create the country's largest tornado outbreak. In just 16 hours, 148 tornadoes hit 11 states killing 330 people and injuring more than 5300. Casualties would have been higher but for the tornado warning that was issued in advance.

The single most violent tornado in the USA took place in Oklahoma City in May 1999. Wind speeds during the tornado were the fastest ever recorded at 512 kph/318 mph and resulted in 36 deaths with $1 billion worth of damage, making it the most costly in the state of Oklahoma.

F-Scale Number	Intensity	Wind Speed
F0	Gale tornado	64–116 kph/40–72 mph
F1	Moderate tornado	117–180 kph/73–112 mph
F2	Significant tornado	181–253 kph/113–157 mph
F3	Severe tornado	254–332 kph/158–206 mph
F4	Devastating tornado	333–418 kph/207–260 mph
F5	Incredible tornado	Over 418 kph/260mph

Opposite: A tornado is defined as a localized and violently destructive windstorm occurring over land, characterized by a funnel-shaped cloud extending towards the ground.

Above right: Most tornadoes occur in the United States, where the large land mass and great temperature contrasts between air masses favour their formation.

Right: Tornadoes may form in the mesocyclone (a strong rotating updraught of warm air), in which case the storm is classified as a tornadic supercell thunderstorm.

Storm Chasers

It is difficult to predict when a tornado will occu[r] since no one is absolutely certain how they form[.] What is predictable, however, is that, if conditio[ns] are right, they can form almost anywhere. Natio[nal] weather services monitor the situations in which tornadoes may form in order to be able to give tornado watch warnings to the media. Meteorolo[gy] offices use the most up-to-date technology to ass[ist] them, including radar and satellite data. Despite t[his] intrepid scientists willingly track weather systems likely to produce tornadoes using specially adapt[ed] vehicles. Storm chasers are involved in the tracki[ng] of tornadoes, collecting valuable data as they do and offering advice on protection from the effec[ts] Their work has provided another set of invaluab[le] data to help the prediction of tornadoes and to provide more accurate warnings.

Storm chasing is thought to have begun d[uring] the 1950s in North Dakota and Minnesota, and i[s] first attributed to such individuals as Roger Jense[n] and David Hoadley, who began photographing a[nd] filming storms recreationally. However, chasers s[oon] became involved in scientific research, and in the early 1970s the Tornado Intercept Project was se[t] up by the University of Oklahoma and the Natio[nal] Severe Storms Laboratory, representing the first organized storm-chasing group that was dedicate[d] to scientific study.

Perhaps the best-known individual workin[g] in the field today is the American Warren Faidley who has been credited with being the first full-ti[me] professional storm chaser, and is regarded as an authority on severe weather.

Most storm chasing takes place on the Gr[eat] Plains of North America, but there are also grou[ps] and individuals dedicated to the pursuit in Aus[tra]lia Israel and much of Europe. Such is their success at tracking tornadoes that it is now possible for t[he] adventurous to take a holiday with storm chasing organizations in the USA.

Left: Storm chasers like to get close to some of natur[e's] most violent weather. Chasers seek out tornadoes, hurricanes, cyclones, waterspouts and thunderstorms as they search for the ultimate storm.

Above: Giant waterspout seen in the Bermuda Triangle area. A waterspout is a funnel-shaped mass of cloud and water, similar to a tornado occurring at sea. It is more a feature of tropical and subtropical waters than of higher latitude seas. An inverted cone-shape cloud descends from a cumulonimbus cloud until it reaches a cone of spray rising from the sea. A column of water is formed between sea and cloud. The spout may be several hundred metres/feet high and can last for half an hour.

Waterspouts

Although we tend to think of tornadoes as a frequent hazard, few people view water spouts in a similar way. They are far more common that many people expect, however, and frequently pose a hazard to those in small craft at sea.

Two types of water spouts form, the more dangerous of which is the tornadic water spout. These often begin overland as conventional tornadoes and drift out to sea. They are very different in formation to the second type, the fair-weather water spout, which are the more common of the two types and are similar to the 'dust devils' or swirls of dust that are sometimes seen on hot days over land in the summer.

Fair-weather water spouts form over open sea or large lakes in the late summer or early autumn when the water temperature is at its highest. If a cool air mass then blows over the warm water the air becomes humid which causes strong convection currents to form. If these are sufficiently strong, water spouts may develop.

At the start a 'dark spot', a small circular area of calm water about 15 metres/50 feet in diameter forms, which is surrounded by rougher water. After a short while a spiral pattern begins to form around the dark spot; wind speeds begin to increase, as does the amount of spray, and a funnel shape of the spray begins to rise off the water surface up to a height of as much as 600 metres/2000 feet. Water spouts tend to last only a short while, perhaps only 20 minutes, and move slowly, at only 30 kilometres/20 miles per hour at most and usually less. Given this, they are inevitably, and thankfully, much less dangerous than tornadoes but remain a real hazard to small ships and recreational sailors.

Above: A 19th century engraving of waterspouts near a ship. Waterspouts form where a tornado touches down on water, instead of land. A tornado is a rapidly rotating funnel of air that can form below certain types of storm clouds. High wind speeds within the rotating funnel can cause immense destruction. Waterspouts are most common over subtropical and tropical waters.

Hailstorms

In May 2000 the worst hailstorm in over 30 years struck parts of Illinois in the USA. During the storm hail ranging from golf ball to tennis ball in size fell. Around 100,000 homes lost power, train services were disrupted and over 100 flights were cancelled. At its worst, the hail was over 8 centimetres/3 inches deep. On another occasion in Hainan province, China, 25 people died and hundreds were injured during a severe hailstorm.

Hailstorms of this magnitude are relatively rare occurrences but the sight of smaller, less damaging hail is much more common. Hailstones are usually only a few millimetres/inches in diameter but can sometimes grow to as much as 15 centimetres/6 inches and weigh more than half a kilogram/1⅛ pounds. Hail of pea or golf ball size are not uncommon in severe storms.

Hail forms in strong thunderstorm clouds where a good proportion of the cloud is below freezing. The growth rate of hail is maximized at about −13°C/9 °F. Hail is most common in midlatitudes during early summer. At this time surface temperatures are warm enough to promote the atmospheric instability associated with strong thunderstorms, but the upper atmosphere is still cool enough to support ice. Despite a much higher frequency of thunderstorms than in the midlatitudes, hail is less common in the tropics. This is because the atmosphere over the tropics tends to be warmer over a much greater depth.

Where condensation nuclei such as dust are present in the atmosphere, supercooled liquid in the clouds can freeze on contact. In clouds containing large numbers of super cooled water droplets, the ice nuclei grow quickly at the expense of liquid droplets. Latent heat released by further freezing may melt the outer coat of the hailstone. The liquid outer shell then allows the hailstone to accumulate other smaller hailstones in addition to supercooled droplets and so it grows. Updrafts in the storm pick up the hailstone during which time it freezes and melts. This process can occur numerous times and as it does so the hail grows increasingly large. Once a hailstone becomes too heavy to be supported by the thunderstorm's updraft it falls out from the cloud. When a hailstone is cut in half a series of layers, such as those of an onion, are revealed. From these rings it is possible to calculate the total number of times the hailstone had travelled to the top of the storm before finally falling to the ground.

Above: A farmer in North Dakota, USA, surveys the damage done to his crop of sunflowers after a hail storm.

Right: Hailstones are particles of ice which fall from cumulonimbus clouds. They are roughly spherical and are often composed of concentric layers of clear and opaque ice. This structure is thought to be the result of the stone travelling up and down within the cloud during its formation. Opaque layers are deposited in the upper, colder parts of the cloud, where the water droplets are small and freeze rapidly, forming a layer of ice with numerous air enclosures. In the warmer, lower parts of the cloud, the water droplets spread over the stone's surface before freezing, so that little air is trapped and the ice is transparent.

Opposite: View of a cumulonimbus hail storm cloud in in the Black Rock Desert, Nevada. The hail forms in cumulonimbus clouds due to the violent convection currents present in these clouds. Particles of rain or snow are drawn vertically up through the cloud, and collide with supercooled water and other particles. These collisions lead to the particles sticking together. When they become too heavy to be supported by the air currents, they fall to earth. Some hailstones can be over 10 centimetres/4 inches across, and cause severe damage to property and crops.

Blizzards and Ice Storms

Winter storms create difficult conditions for everyone who experiences them. In 1993 a massive storm dropped snow in blizzard conditions, which affected half the population of the USA over 26 states as well as southern parts of Canada and even as far south as Mexico: 270 people died and a further 48 were missing at sea. Snow lay as deep as 4.5 metres/15 feet and up to 15 metres/50 feet fell, including as much snow as normally falls in Alabama in an entire winter. No wonder it was called the storm of the century.

So what is a blizzard? Different definitions exist but one in common use states that visibility must be down to 400 metres/0.25 mile for at least four consecutive hours and must include snow or ice as precipitation and wind speeds must be at least 56 kilometres/35 miles per hour – 7 on the Beaufort Scale. Blizzards can occur when polar cyclones develop. An extreme version of this is a whiteout where downdraughts bring snow, making it virtually impossible to distinguish between the ground and the air.

The worst ice storm to affect the Great Lakes area of the USA and Canada took place over a few days in early January 1998. The Canadian side of the border experienced 80 hours of freezing rain, twice as much as the annual average. This led to 28 deaths, mainly from hypothermia, 945 injuries and 4 million homes left without power, forcing 600,000 people to leave their homes in Ontario, Quebec and New Brunswick. The environment suffered too, with millions of trees lost because of damage from the cold and ice. By the summer of that year over $1 billion of insurance claims had been filed and it is believed that the final cost of the storm was almost $5.5 billion.

Left and above: Ice storms can be extremely destructive when the glaze (a layer of ice that coats objects) snaps trees, utility poles and electricity pylons.

Opposite: The combination of heavy falls of snow, very cold temperatures, zero visibility, deep snow drifts and potentially lethal wind chill can make it extremely unpleasant to move around in a blizzard or ice storm.

Below: An ice storm is a severe weather condition characterized by falling freezing precipitation. Such a storm forms a glaze on objects, often creating hazardous travel conditions and utility problems.

Avalanches

In February 1999 in Galtuer, Austria a series of huge avalanches killed 38 people. Massive snowfall – up to 40 centimetres/16 inches in one day – was made unstable by strong winds. At the end of the year the same region was hit by yet more avalanches which killed a further 15 people. Between 1998 and the start of 2000 Austria recorded 56 avalanches which killed 50 people and injured a further 70. Austria and the other Alpine countries of France, Switzerland and Italy are those which are hit most frequently by avalanches, but the events of 1999 were nevertheless highly unusual.

Avalanches are a seasonal hazard which occur when a large amount of snow (or rock) slides down a mountainside. When snowfall is heavy and a significant build-up occurs, an avalanche is likely to occur.

There are various factors that may come togther to increase the likelihood of the ocurrence of avalanches. Anything more than a gentle wind can to contribute to the build-up of snow on downwind slopes. This can make the snow unstable and may precipitate avalanches.

The most obvious factor is, of course, heavy snowfall. This makes the snow on the slope unstable because of the extra weight and also because the snow has not had time to bond; the first 24 hours after a storm are the most critical. Rain can also play a part as it too adds extra weight and may help lubricate lower layers of snow.

The steeper the slope, the greater the risk. Slope aspect may also contribute to the avalanche hazard as a southerly aspect will be warmer.

There are two types of avalanche, loose snow and slab avalanches. The former are generally fairly minor and rarely deadly, they form when powder snow falls on the mountain and simply cascades down the slopes. The slab avalanche, however, is far more dangerous and is the type responsible for the deaths in Galtuer. They can carry trees, rocks and other debris with them as they rush down the slopes, making them highly dangerous. Weather plays a significant determining role in the formation of slab avalanches. The most important elements are heating by solar radiation, radiational cooling, temperature gradients within the layers of snow, and snowfall type and amount. If the temperature is high enough for alternate sequences of freezing and thawing to take place then the melting and re-freezing of water in the snow strengthens the snow during freezing but weakens it during melting. In spring, as temperatures rise above freezing point, this can weaken the whole structure causing the bond between the top slab and the lower layers to break and so trigger an avalanche.

Left: Avalanches occur when large amounts of snow and ice become loose and flow down a mountain under the force of gravity.

Dust Storms

Dust or sand storms are common occurrences in arid or semi-arid areas of the world such as deserts. When the land is heated by intense solar radiation, convection takes place; in other words the air rises, creating an area of low pressure. It is replaced by cooler air rushing in from an area of higher pressure, and this is wind.

The inrush of wind can be fast enough to move sand dunes in desert areas like the Sahara and can create problems for travel, often obscuring roads and creating very low visibility conditions. If the topsoil is light and uncultivated the wind may blow it away and deposit it where it is not wanted, creating a nuisance. Dust and sand storms can move material many thousands of kilometres/miles. At times the leading edge of a sand or dust storm may appear as a solid wall reaching as high as 1520 metres/5000 feet.

In November 1991 in California a dust storm formed over Interstate Highway 5. It picked up dust in a storm where wind speeds exceeded 120 kilometres/75 miles per hour, causing visibility to fall to less than half a car's length. Seventeen people were killed and 151 injured in the resulting 164-car pile-up.

A local wind, the simoom or simoon is common in the Khartoum region of Sudan. It is largely responsible for the atmospheric dust found over Europe. This is also found on the seafloor many thousands of kilometres/miles from its point of origin.

In addition to the danger they pose, dust storms can be a problem for farmers. Drought and wind contribute to the problem as does poor cultivation or over-grazing of topsoil. During the 1930s in the USA the 'dustbowl' conditions of the mid-western states led to the loss of topsoil from some areas and its deposition, often tens of metres/feet deep, in others. These storms were sometimes referred to as 'black blizzards' as the dust and sand was blown into the atmosphere, sometimes so thickly that it blocked out the sun for several days at a time. Many farmers were forced off their land as a result of the severe economic hardship that resulted.

Right: Sahara Desert in Chad seen from space. A sandstorm is seen at upper centre (white) over the Djourab region. The winds in this region blow predominantly towards the south-west, towards the horizon in this image. They are channelled between two highland areas (dark), the Tibesti Mountains (centre right) and the Ennedi Massif (centre left). These winds have created large streak features on the plains. The Tibesti Mountains are the highest point in the Sahara Desert.

Opposite below: Sossusvlie Dunes, Namib Desert, Namibia. Sandstorms from the Namib Desert can carry dust far into the South Atlantic. Long periods of heat coupled with little rain help to produce the fine sand of desert dunes.

Below: Women caught in a sand storm in Mali.

Monsoons

Over half of the world's population relies on the rains brought by the monsoon rains of Asia, Africa and Australia. It is the so-called 'monsoon subcontinent' of south and south-east Asia – India, Pakistan and Bangladesh – that we tend to associate most with this rainfall phenomenon. More than 75 per cent of India's annual rainfall occurs during the monsoon. In the town of Cherrapundi in north east India the average rainfall in December is only 13 millimetres/½ inch whereas in June its average rainfall increases to a staggering 2695 millimetres/106 inches. The total amount of rain that falls here is over 12,200 millimetres/500 inches each year.

The word monsoon is believed to originate from the Arabic word 'mausim', which means season, because of the seasonal changes in wind direction in the Tropics. It is applied to the two main seasons in the Tropics which are caused by variations in air pressure in Asia.

During the Northern Hemisphere winter the high pressure system over Siberia intensifies, causing strong north-easterly winds which blow towards the Equator. They remain dry until they blow across the South China Sea and then become the north-eastly monsoon of south-east Asia. The second cause is the intense low that forms over central Asia which draws in the south-easterly Trade winds that deflect towards the south-west as they cross the Equator. This is responsible for the heavy rainfall over southern Asia.

Monsoon Effects

We tend to assume that the monsoon is a regular, reliable source of rainfall but this is not always the case. If the rainfall is late or less than expected it can lead to crop failure and starvation. In India, a country where the population is already 1 billion and is predicted to exceed 1.4 billion by 2025, the rainfall is vital and even minor variations can be catastrophic.

The rainfall is the most striking aspect of a monsoon but sometimes has the unwanted side effect of flooding. This is short-term but nonetheless highly disruptive, causing immense damage and financial loss to countries that find it difficult to cope at the best of times.

Modern forecasting methods allow governments to plan for the onset of the monsoon to reduce these effects. It is now possible to predict the arrival of the monsoon to within two or three days.

Above: Rickshaws battle through deep water in the streets during the monsoon season in India. The Tropical Rainfall Measuring Mission (TRRM) records rainfall over Mumbai and Maharashtra State, India. The heaviest rainfall ever recorded in Indian history was on 26 July 2005; 944 millimetres/37 inches of rain fell in a 24-hour period. This monsoon rain, which was combined with high tides, caused massive flooding and damage. It is thought that up to 1000 people were killed.

Left: Indian labourers plant paddy cuttings in a field near Amritsar, in India's northwestern state of Punjab. With the arrival of the monsoon rains, fields that have been barren for months become fertile again and replenish water supplies for the dry season ahead.

Right: The monsoon areas of the world.

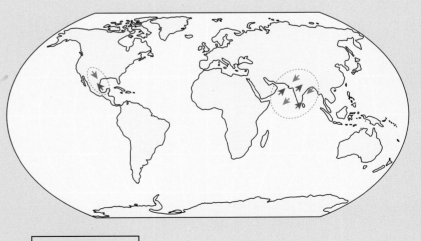

Monsoon Areas

◄ Winter monsoons
► Summer monsoons

Flooding

For many people who live in a river valley the risk of flooding taking place is common and may be responsible for damage to property and disruption to travel. In more extreme cases, however, flooding can be catastrophic and may lead to enforced evacuation of homes, loss of power and numerous deaths. Flooding can also lead to landslides and even avalanches, adding to the overall risk posed.

There are, broadly speaking, two types of flood – the river or mainstream flood and the flash flood. The river or mainstream flood is usually caused by heavy rainfall over an extended period of time, perhaps as a frontal system passes, or as a result of spring snow melt. In contrast, a flash flood is usually caused by a short period of intense rainfall such as during a slow-moving thunderstorm.

Above: Flooding on the Mississippi River, 1912. The Mississippi, one of the longest rivers in the United States, flows over 3700 kilometres/2,300 miles north to south across the country. It has the third largest drainage basin in the world, and heavy rainfall can cause massive flooding further downstream and kill large numbers of livestock. Some of the worst floods in America have occurred on the Mississippi River, including the Great Mississippi Flood of 1927.

Left: A flash flood is caused by a sudden heavy downpour of rain from a storm.

About 40 per cent of all casualties in natural disasters occur in floods: in the USA floods are the second biggest killer, responsible for an average of 84 deaths between 1984 and 2003. In only 15 centimetres/6 inches of fast flowing water a person can be knocked off his or her feet and in similar conditions with 60 centimetres/24 inches of water it is possible to see cars washed away.

Flooding is sometimes worse in urban areas where summer thunderstorms are common and the ground is tarmac. This material is impermeable – does not allow water to drain through it – and so cannot absorb the extra water that has to run off the surface. Urban areas are deliberately designed to ensure water is drained away swiftly into rivers and this inevitably increases the flood risk. Flooding is, of course, a natural event on rivers. The annual flooding of the Nile was a critical part of the way people used the land, because it provided the silt that created the natural fertility of the soils, which the population depended upon. The Mississippi in the USA also regularly flooded but this was never really a problem until human settlement along the flood plain developed from the late 19th century. Today huge sums of money are spent on protecting, not always successfully, the settlements that have grown up there from the risk of flooding.

The most severe floods are rated according to severity by the frequency with which they take place. A one in 100 year flood will be of far greater intensity that a one in five year flood, cover a far larger area and have far greater impact on property and lives. This will have an effect on where – and what – flood defences may be put in place. Typical flood defences include 'soft engineering' schemes such as earth embankments and tree planting or more costly 'hard engineering' such as dams and flood barriers. It may not be cost-effective to protect against a one in 100 year flood given its relative infrequency and the likely high cost of the inevitable engineering it will involve. Instead it may be more sensible to consider flood plain zoning, preventing vulnerable land uses from being allowed on at-risk areas.

Right: In Europe, floods are the most common natural disasters. Between 1975 and 2001, 238 major floods were recorded in the region. Since 1990 about 2000 people have died and some 400,000 have been left homeless.

Above: Aerial photograph of the Thames flood barrier with the gates closed. The gates are shut when there is a high risk of tidal flooding, or for maintenance. Closing the barrier seals off part of the upper Thames with a continuous steel wall spanning 520 metres/1700 feet across the river.

Prague and Europe 2002

The summer of 2002 saw widespread flooding over much of central and western Europe, from Romania to Russia. The city that was perhaps most badly affected and was certainly most in the news was the Czech capital, Prague. In August of that year heavy rain fell from a slow-moving frontal system which had also affected much of the rest of Europe. The river Vltava rose ominously, forcing the country's president to enforce an evacuation of 40,000 people. Despite this the flood defences were breached, leading to many road, rail and tram links being cut and causing significant damage to the city's architectural wonders. At one point there was a fear that the famous landmark, the Charles Bridge, would be badly damaged. Many other towns and cities in the country were equally badly disrupted as rivers flooded, leading to 10 deaths and hundreds of thousands of people being made homeless.

At the same time in Austria the worst floods for a century claimed a similar number of lives, as did the events in Romania, but it was in the Black Sea area of Russia that most lives were lost. Here 58 people died and over 1500 were evacuated as the floodwaters flowed down the Danube.

It has not been possible to fully explain why the floods were so bad that summer but a body of evidence now hints at the effect of El Niño, the warm water current that is believed to have a significant impact on global weather patterns.

Below: 14 August 2002: Czech soldiers and volunteers build a sandbag wall to try to hold back the rising waters of the Vltava river in Prague. Flood waters began to spill into Prague's historic old town as emergency workers raced to reinforce the sandbag walls protecting centuries-old architecture. as well as organizing a large number of evacuations.

Flash Flooding

Flash floods are by their very nature unpredictable and potentially deadly. A flash flood occurs within a few hours of the rainfall event. The rainfall is usually a very intense and short burst of rain over a relatively small area after a heavy storm such as a summer thunderstorm. Humid air will rise quickly and if it remains relatively stationary it will drop all its rain over a small area of land. If humid air rises over mountains and is kept in a small area by the wind it will also drop most of its rain over the hills, forcing it to flow down hillsides into the valley below. When the ground is baked hard by the sun any rain that falls will not be absorbed by the Earth. Instead it will be forced to run off over the surface and make its way into rivers and canyons. Flash floods can also occur in desert areas where the dry, sun-baked soil cannot absorb the rain that falls and this can turn a dry river bed or wadi into a raging torrent of water in minutes. In the deserts of North America more people die as a result of drowning in flash floods than from thirst or heat exhaustion. Water in flash floods often moves at incredible speeds catching people unawares and can also lead to mudslides.

One example occurred in the USA in June 1990 in Shadyside, Ohio. Here 10 centimetres/4 inches of rain fell in less than 2 hours creating a 9 metre/30 feet high wall of water which then flowed through parts of the town. Twenty-six people died and it was estimated that between $6 and 8 million worth of damage was caused.

Above: Mozambicans are trapped after heavy rains in Xai-Xai province in 2000. Flooding has forced an estimated 800,000 Mozambicans to find temporary refuge and left many dead across southern Africa. Flooding threatened the fertile farming land of the Limpompo Valley and the regional capital and the threat of waterborne disease loomed.

Right: The Bisagno River rages under a bridge in Genoa, Italy.

Even more devastating however, were the flash floods that occurred in China in 1998 and Mozambique in 2000. In the summer of 1998, torrential rainfall in July and August caused the massive Yangtze River to flood. Collapsing levees washed away recently constructed dykes and caused huge landslides and mudflows. The problem was apparently worsened by poor land use: deforestation, and the drainage of water-absorbing wetlands for farming. Some nine million hectares of arable farmland were inundated, completely ruining the crops that were being grown, whilst almost three million houses were destroyed, 14 million people were made homeless, and around three thousand lost their lives. Overall, it was estimated that about 240 million people, or a fifth of China's population, was adversely affected.

Flash flooding also caused severe destruction and serious loss of life in Mozambique in 2000 when heavy rains caused the banks of the Limpopo River to burst. The capital of Mozambique, Maputo was flooded, as were the Limpopo, Save and Zambezi valleys. The disaster was then exacerbated by tropical cyclone Eline, which struck just north of the affected areas a few days later. Over 140,000 hectares/345,954 acres of crops were destroyed, 20 thousand cattle were lost, and about half a million people were made homeless. Additionally, around 800 people lost their lives, and many more were affected by diseases such as dysentery on account of the destruction of the sanitation infrastructure and a lack of clean drinking water.

Above: A Chinese couple collect sweetcorn from their field that has been damaged by floods.

Opposite: False-colour satellite image of floods caused by monsoon rains in Bangladesh (centre left) and India (far left, centre right). Rivers have burst their banks and the terrain surrounding the Ganges River (centre left) and Brahmaputra River (upper right) is waterlogged. Standing water is visible, particularly in Dhaka (centre), the capital city of Bangladesh.

Floods in Bangladesh

Perhaps the area most associated with serious flooding is the small country of Bangladesh in Asia. Its physical geography lies at the heart of the causes of flooding in the region.

Bangladesh is situated at the head of the Bay of Bengal where it narrows as it reaches the Ganges Delta. To the north lies the Himalaya range with its permanent ice and snow fields. Some 150 rivers criss-cross the country including the Ganges itself and two major tributaries, the Brahmaputra and the Meghna. Over 50 per cent of the country is less than 5 metres/16 feet above sea level and is affected every year by torrential rain from the annual monsoon.

With a population of 160 million, Bangladesh is one of the world's most densely populated countries and also one of the poorest. Many people are forced to live subsistence lives on the fragile and dynamic delta region, which naturally floods, causing the position of the innumerable islands and distributaries it comprises to change.

The annual floods, then, are a necessary evil; they provide the water required for irrigation of crops, such as rice and millet, and add fertile silt to the soil. As a rule, their effects are not catastrophic but can bring these important benefits.

There is, however, some evidence to suggest that human activity in the Himalayas, such as deforestation of the headwaters of the major rivers, is adding to the problem of flooding downstream. Removing the trees prevents take-up of water by vegetation and allows the rivers to more easily erode the banks. This adds extra silt to the river channels downstream and reduces their capacity to hold any additional water during potential floods.

In addition to all this, Bangladesh's low-lying land makes it particularly vulnerable to the storm surges associated with tropical cyclones – hurricanes – that frequently affect the area. This water is, of course, salty, and has the unwanted effect of making the soil increasingly saline and unusable.

In 1991 a major flood, a combination of events, led to catastrophe. A storm surge 6 metres/20 feet high inundated the land and contributed to 125,000 deaths. Many of these were due to drowning but a significant number were caused by the subsequent hazards associated with homelessness and disease.

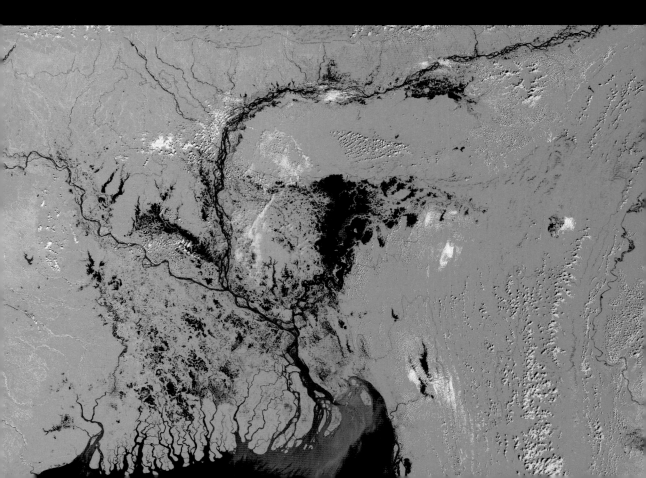

Drought

Although there are over 100 different definintions, a drought may be best defined as a period of time when there is significantly less rainfall than normal. Being more precise than that is difficult; in the USA a drought is a period of a month or more with less than 30 per cent of normal rainfall but in the UK it is possible to experience a drought when a period of only three weeks of no rainfall occurs.

In some countries drought is a way of life, a commonly occurring event which the population living in the area have learned to manage over the centuries. For others, periodic droughts cause chaos. The effects can be devastating, with failed crops causing significant economic difficulties that cost billions of pounds, even famine, deaths and malnutrition-related disease. The impact on the natural environment is equally devastating with soil erosion, and loss of native vegetation and animal habitats. Frequent, persistent drought can lead to the ideal conditions for bush fires and dust storms.

The British drought of 1976 is believed to have been caused by the jet stream being forced further north than normal by a so-called 'blocking high', a particularly well-developed high pressure system which sat over much of north-west Europe for several months. The low pressure systems moving east were forced further north by high pressure and pulled along by the jet stream. This compounded the problem which had begun during the very dry summer, autumn and winter of 1975. The 1976

Left: A Jeffrey Pine, Yosemite National Park, California affected by lack of water.

Opposite above: Parched mud, Orange River, Floodpan, South Africa. Some areas receive less rain than others.

Opposite below: Aral Sea boat stranded amongst dunes. The Aral Sea, inland between Kazakhstan and Uzbekistan, was supplied by rivers, but in the 1930s these were diverted to irrigate crops, notably cotton. As less water flowed to the sea, it began shrinking through evaporation. It has halved in area since the 1960s, and its salinity has increased dramatically, leaving ghost towns that were once thriving fishing communities and damaging the ecosystem.

drought began in earnest in May: virtually no rain fell until August, then ten days of almost continuous rainfall followed, ironically causing flooding. The government appointed a Minister for Drought and passed a Drought Act, which allowed for extreme emergency measures such as standpipes to be set up in streets. Lost crops to the value of £500 million were blamed on the drought that year.

The Sahel region south of the Sahara Desert in Africa normally has unreliable rainfall that can be managed by the small native population who have adapted strategies over the centuries. When this turned into a long-term drought during the 1970s and 1980s, however, the effects in countries such as Ethiopia and Kenya were catastrophic. Deaths ran into the thousands and many tens of thousands were temporarily or permanently displaced.

In the 1930s the great Dust Bowl drought severely affected much of the USA. It was particularly severe in 1934, 1936 and 1939–40 but some regions experienced eight years of continuous drought. The lack of rain was exacerbated by the poor agricultural practices of the time. The grasslands of the Great Plains had been deeply ploughed and, when rainfall was sufficient, produced good yields of cereals. Once the drought really hit, however, nothing would grow and the soil was left bare and exposed to the wind. Many farmers had no alternative but to leave their farms and seek a new life elsewhere such as California. The effects of this drought led to profound political and social changes as well as significant environmental damage.

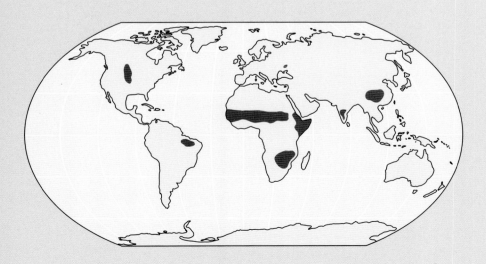

Left: Areas at risk of drought are marked in red. In some countries, drought is a way of life and their populations have learned to adapt, but prolonged or unexpected drought can cause chaos.

Forest and Bush Fires

These are a natural phenomenon that are most common in Australia but which occur wherever there is wood, leaves and scrubby vegetation that can burn. They are sometimes called wild fires as well as bush or forest fires.

The causes are most commonly associated with drought and lightning strikes and sometimes human action. Where the climate is moist enough to allow vegetation growth, but has hot and dry periods that allow for dried-out branches and leaves, there is the potential for fire. This material is highly flammable so following a drought, and when the winds are strong, the smallest spark can ignite a fire which can spread rapidly through an area.

One of the earliest records of forest fire has been found in fossil records from the Devonian period, about 365 million years ago. It was then, and remains today, a natural part of many ecosystems, and plants and some animals have evolved to adapt to those conditions. Eucalypts, common in Australia, contain flammable oil in their leaves, which means they encourage fire and this allows the plants to out-compete other non-fire-retardant species. When the water content of the soil is between 30 and 100 per cent the evaporation of water in plants is balanced by the water in the soil. Below this threshold the plants dry out and, under stress, release ethylene, a flammable gas. This is inevitable after a long dry period, as are the fires which ensue.

Fires are inevitably more common in summer where the conditions are right. The old wood, sometimes called 'slash', which has fallen to the ground is ideal fuel. In Los Alerces National Park, Argentina, this slash is routinely cleared from the areas where humans visit because the high risk of lightning strikes can lead to huge fires.

Many people believe that forest fires are more common today. This may not be true: the extension of urban areas into rural, undeveloped areas naturally prone to fires means more properties are now at risk than in the past. This has led to considerable destruction of property and loss of life in areas such as Australia and California, USA. Fire tends to move fastest up a slope so new housing built at the top of hills to take advantage of views is very vulnerable. Such fires are an increasing hazard in the USA, particularly in the west. Because of increased building in vulnerable areas small fires are forbidden, which means the build-up of flammable material can reach a critical point and the resulting fires are enormous and particularly damaging. In 2004 over 3.25 million hectares/8 million acres of land burned across 40 states. In 1988 a fire in Yellowstone National Park destroyed 490,000 hectares/1.2 million acres.

In Australia during September 2006 over 50 separate fires engulfed hundreds of homes and destroyed thousands of acres of park and farm land as 100 kilometres/60 miles per hour winds blew across the area on the outskirts of Sydney. Following a prolonged drought, experiencing such a large-scale bush fire so early in the southern hemisphere spring did not bode well for the rest of the summer.

Wherever fires occur the effects are the same: property is destroyed, people die and the ecosystem is damaged. Sometimes landslides occur as there is no vegetation to hold the soil in place, and the smoke and ash in the atmosphere reduces visibility and damages air quality.

Left: Smoke rising from the flames of a bush fire in California, USA. Most fires are the result of drought and lightning strikes. However, they are increasingly caused by human activity, whether deliberate burnoffs, unextinguished campfires or discarded cigarette butts. Small blazes can get out of hand very quickly and turn into raging infernos.

WEATHER RECORDS

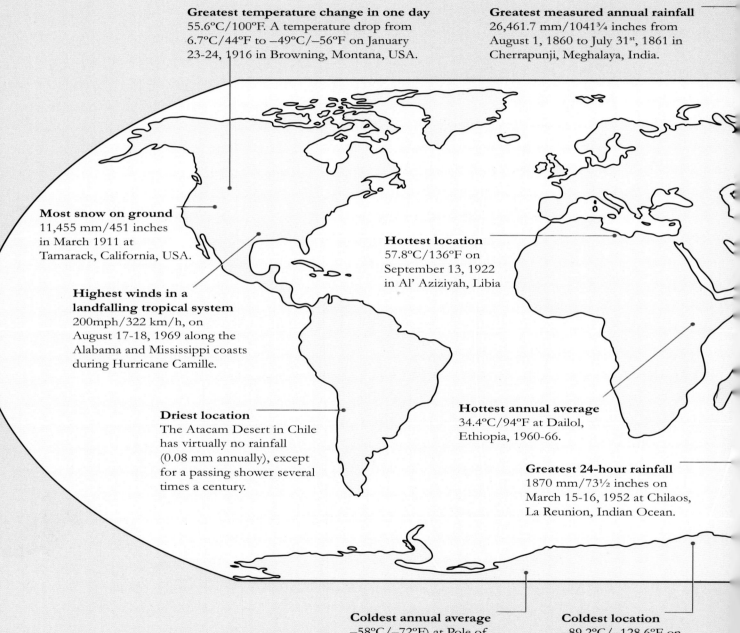

Greatest temperature change in one day
55.6°C/100°F. A temperature drop from
6.7°C/44°F to −49°C/−56°F on January
23-24, 1916 in Browning, Montana, USA.

Greatest measured annual rainfall
26,461.7 mm/1041¾ inches from
August 1, 1860 to July 31st, 1861 in
Cherrapunji, Meghalaya, India.

Most snow on ground
11,455 mm/451 inches
in March 1911 at
Tamarack, California, USA.

Hottest location
57.8°C/136°F on
September 13, 1922
in Al' Aziziyah, Libia

**Highest winds in a
landfalling tropical system**
200mph/322 km/h, on
August 17-18, 1969 along the
Alabama and Mississippi coasts
during Hurricane Camille.

Driest location
The Atacam Desert in Chile
has virtually no rainfall
(0.08 mm annually), except
for a passing shower several
times a century.

Hottest annual average
34.4°C/94°F at Dailol,
Ethiopia, 1960-66.

Greatest 24-hour rainfall
1870 mm/73½ inches on
March 15-16, 1952 at Chilaos,
La Reunion, Indian Ocean.

Coldest annual average
−58°C/−72°F) at Pole of
Inaccessibility, Antarctica.

Coldest location
−89.2°C/−128.6°F on
July 21st 1983 at Vostok
station, Antarctica.

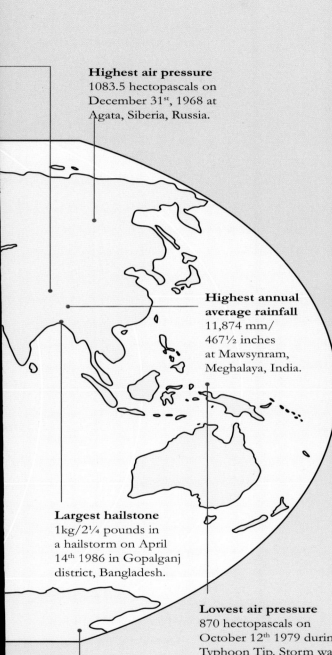

Highest air pressure
1083.5 hectopascals on
December 31ˢᵗ, 1968 at
Agata, Siberia, Russia.

**Highest annual
average rainfall**
11,874 mm/
467½ inches
at Mawsynram,
Meghalaya, India.

Largest hailstone
1kg/2¼ pounds in
a hailstorm on April
14ᵗʰ 1986 in Gopalganj
district, Bangladesh.

Lowest air pressure
870 hectopascals on
October 12ᵗʰ 1979 during
Typhoon Tip. Storm was
300 miles/483 km west
of Guam, Pacific Ocean.

Windiest location
Winds reaching 200mph/
322km/h at Commonwealth
Bay, George V Coast, Antarctica.

Wind

Windiest place at high altitude
The USA holds this record on Mount Washington in
New Hampshire at an altitude of 1918 metres/6288
feet. On 12 March 1934 the wind speed reached
371.75 kilometres/231 miles per hour.

Windiest place at low altitude
The fastest surface wind speed found at low
altitude occurred in Thule, Greenland on 8 March
1972. At the local USAF base they recorded
333 kilometres/207 miles per hour.

Tornadoes – the greatest number per area
Although most people would assume that the United
States would have most tornadoes in this category
this is not the case. Instead, that honour falls to the
UK. In the USA there is one tornado per 8663
square kilometres/2856 square miles. In the UK,
however, the number is one per 7397 square
kilometres/2856 square miles.

Tornadoes – the greatest number in 24 hours
Not surprisingly this record is held by the infamous
Tornado Alley area of the USA. Here, on the 3 and 4
April 1974 a staggering 148 tornadoes were recorded.

Tornadoes – the largest
Once again, Tornado Alley in the USA holds this
record. Mulhall in north Oklahoma experienced
a huge tornado on 2 May 1999. This, one of
70 which occurred that day, was 1600 metres/5,250
feet in diameter and caused 40 deaths.

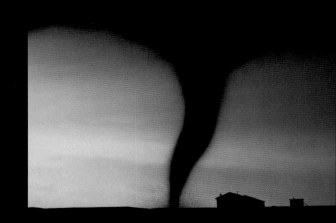

Above: Black lava rocks on Hawaii. Despite their difference in latitude, Alaska and Hawaii share the same heat record. In Pahala, Hawaii the temperature on 27 April 1931 was 38°C/100°F.

Opposite above: Mount Mckinley, Alaska. Alaska's highest recorded temperature is also 38°C/100°F on 27 June 1915 at Fort Yukon.

Opposite centre: The worst death toll from a lightning strike occurred when a Boeing 707–121 jet liner was struck by lightning near Elkton, Maryland, USA on 8 December 1963. Eighty-one people were killed when the lightning strike ignited fuel vapour causing a mid-air explosion.

Opposite below: The world's highest temperature was recorded at Al Azizayah, in the Libyan Sahara on 13 September 1922.

Temperatures

HIGHEST

The highest temperatures ever recorded by continent:

Africa: El Azizia in Libya at an elevation of 112 metres /367 feet experienced the world's highest temperature of (57.7°C/136°F) on 13 September 1922.

North America: The Greenland Ranch in Death Valley, California, USA, is 54 metres/178 feet below sea level and it experienced a temperature which reached 56.6°C/134°F on the 10 July in 1913.

Asia: The highest temperature recorded in Asia occurred in Israel at Tirat Tsvi on 22nd June 1942. Like Death Valley this area is below sea level, some 220 metres/722 feet. The temperature experienced here was 53.9°C/129°F.

Australia: Cloncurry, Queensland, is at 190 metres/622 feet above sea level. It experienced a temperature of 53°C/128°F on 16 January 1889. Given the date there maybe some doubt as to the validity of this figure as the instruments available then were different to the standard used to day.

Europe: 4 August 1881 saw the hottest ever temperature recorded in Europe in Seville, southern Spain. A temperature of 50°C/122°F was recorded in this Andalusian city.

South America: In 1905, Rivadavia in Argentina recorded 48.8°C/120°F, the hottest ever in South America.

Oceania: On 29 April 1912, a record 42.2°C/108°F was recorded at Tuguegarao in the Philippines.

Antarctica: The somewhat modest 15°C/59°F is the highest ever recorded in this continent on 5 January 1974.

LOWEST

The lowest temperatures ever recorded by continent.

Antarctica: Vostock, situated at an elevation of 3413 metres /11,220 feet, experienced a staggering −89.2°C/−129°F on 21 July 1983.

Asia: The honour of second place is shared by two stations in Asian Russia – Siberia. The settlements of Oimekon and Verkhoyansk both experienced −68°C/−90°F, the former on 6 February 1933 and the latter on 7 February 1892.

Greenland: At Northice on 9 January 1954 at an elevation of 2343 metres/7687 feet a temperature of −66°C/−87°F occurred.

North America: The Yukon, Canada is the location of the continent's lowest temperature, which was −63°C/−81.4°F on 3 February 1847.

Europe: On a January day (the precise date is unknown) European Russia experience its, and the continent's, lowest temperature of −55°C/−67°F at Ust'Shchugor.

South America: Argentina's and the continent's lowest temperature of –32.8°C/–27°F occurred at Sarmiento on 1 January 1907.

Africa: At an elevation of 1635 metres/5364 feet, Ifrane in Morocco experienced a temperature of –23.8°C/–11°F on 11 February 1935.

Australia: Australia is not a part of the world normally associated with very cold temperatures but Charlotte Pass, New South Wales, at an elevation of 1755 metres/5758 feet experienced –23°C/–9.4°F on 29 June 1994.

Oceania: In May 1979 the Mauna Kea Observatory on Hawaii, at an elevation of 4198 metres/ 13,773 feet, experienced the continent's lowest temperature –11°C/12°F.

Greatest Temperature Range in One Day

At an elevation of 1170 metres/3840 feet near Glacier National Park, the town of Browning, Montana in the USA experienced an enormous range of 42°C/110°F. Over the 23 and 24 January 1916 the temperature varied from –7°C/44°F to –49°C/–56°F.

Greatest Annual Temperature Range

This was recorded in Verkhoyansk, Russia, where the temperature range is 105°C/188.4°F. The maximum temperature recorded there was 37°C/98°F with the minimum only –68°C/–90.4°F.

Precipitation

DRIEST PLACE (annual averages)

South America: This is the Atacama Desert in Chile. Between 1964 and 2001 the annual average rainfall total at Quillaga was less than 0.5 millimetres/¹⁄₄₈ inch. At Arica, also in Chile, there is annually an average of 0.8 millimetres/¹⁄₃₀ inch.

Africa: At Wadi Halfa, Sudan, less than 13 millimetres/½ inch of rain falls annually, a record that has stood for almost 40 years.

Antarctica: At the Amundsen-Scott South Pole station less than 20 millimetres/⅞ inch falls each year on average.

North America: Mexico holds this record; at Batagues only 31 millimetres/1¼ inches of rain falls on average every year.

Asia: Aden, in Yemen, holds Africa's record for the lowest annual average rainfall. At just 46 millimetres/1¾ inches, the record has been held for 50 years.

Australia: Mulka, South Australia holds this continent's record of 103 millimetres/4¼ inches.

Europe: The city of Astrakhan in Russia experiences only 163 millimetres/6½ inches of rain every year.

Greatest number of consecutive sunny days: Between 9 February 1967 and 17 March 1969, St Petersburg in Florida experienced 768 consecutive days of sunshine.

Greatest hours of sunshine: Sunny Yuma, Arizona receives an average of 4,055 of 4,456 possible hours of sunshine each year, 91 per cent of the total possible.

Most rainy days: At 1,568 metres/5144 feet above sea level, Kauai in Hawaii was always likely to experience a great deal of rain but there are usually only two days when it doesn't rain. On average this area receives up to 350 rainy days each year and these bring almost 12,700 millimetres/500 inches of rain each year.

Oceania: 227 millimetres/9 inches of rain is the expected annual average on this continent. The record is held by Puako, Hawaii.

Highest monthly rainfall:
In July 1861, Cherrapunji in India received 9300 millimetres/336 inches of rain.

WETTEST PLACE (annual averages)
South America: 13,299 millimetres/523½ inches of rain falls on average each year in Lloro, Colombia. This is an estimate and, if accurate, beats the former official record of 8992 millimetres/354 inches at Quibdo, also in Colombia, and only a few miles away but at a lower altitude. If this is correct then it would put it in fifth position by continent. Whichever is the final winner, either is a remarkable total.

Asia: This continent's highest total is believed to be Mawsynram, India. Its high elevation, 1400 metres/4597 feet, may have helped it to reach its total average of 11,684 millimetres/467½ inches each year.

Oceania: Hawaii, once again, holds this record with Mount Waialeale receiving 11,684 millimetres/460 inches each year.

Africa: The country of Cameroon holds his continent's record. At Debundscha 19,827 millimetres/405 inches of rain is the annual average total.

Australia: At an altitude of 1555 metres/ 5102 feet above sea level, Bellenden Ker in tropical Queensland holds Australia's record. A hefty average of 8636 millimetres/340 inches of rainfall occurs here each year.

North America: British Columbia in Canada holds North America's record. At Henderson Lake the annual average is 6502 millimetres/256 inches.

Europe: At Crkvica in Bosnia Herzegovina, at an altitude of 1017 metres/3337 feet, the annual average is 4648 millimetres/183 inches. This record has been held for 22 years.

It is worth noting that the figures quoted here are based on many different sources of information and may differ from others. This may be accounted for by variations in measurement practices and procedures and periods of measurement being different at depending on location and time.

Snowfall and Hailstones

Highest Annual Amount of Snowfall:
The volcano Mount Rainier holds this record. Between February 1971 and February 1972 a total of 31,102 millimetres/4224 inches of snow fell.

Highest Amount of Snowfall in One Day:
The world record for the greatest amount of snow fall on one day was an astonishing 192 centimetres/ 76 inches at Silver Lake, Colorado, USA on 15 April 1921. In Canada the highest amount fell on 11 February 1999 in Tahtsa Lake, British Columbia, Canada. Here 145 centimetres/57 inches fell in a single day.

Highest Amount of Snow from a Single Snowstorm:
Another American volcano holds this record. In 1959, during a storm which occurred between 13 and 19 February, at Ski Bowl on Mount Shasta, 480 centimetres/189 inches of snow fell.

Left: A large hailstone. However, the biggest hailstone in American history landed in Aurora, Nebraska during a violent thunderstorm on 22 June 2003. The piece of ice was almost 18 centimetres/7 inches in diameter with a circumference of 47.5 centimetres/19 inches. Hailstones grow larger as they circulate until they are either too heavy to be supported by the cloud or are thrown out of the updraught area.

Opposite above: Arctic landscape, Lancaster Sound, Nunavut, Canada. In recent years the effects of global warming have caused sea ice to melt earlier in the season and glaciers to recede so it is difficult to predict what the nature of future weather records will be.

Opposite below: Pedestrians battle against snow and wind. The most snow to fall on the ground was recorded on 11 March 1911, when 11,455 millimetres/451 inches fell in Tamarack, California.

Largest Snowflake:

According to records from January 1887 an enormous snowflake fell on Fort Keogh in Montana, USA. This flake was reported to have been 38 centimetres/15 inches in diameter and 20 centimetres/8 inches thick. This one alone would have made several very useful snowballs!

Heaviest Hailstone:

The world's heaviest hailstone fell in Bangladesh in April 1986. This weighed 1 kilogram/2¼ pounds and fell in a storm which killed 92 people. In 1888 in Moradabad cricket or baseball-sized hailstones fell killing dozens of people and up to 1600 cattle.

UNUSUAL
WEATHER OPTICALS

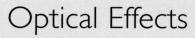

Optical Effects

There can be few people who do not recall learning the colours of the rainbow in school but how many of us can remember exactly why there are these colours and why a rainbow exists at all?

Sunlight consists of a mixture of colours from across the spectrum. We can see this most clearly when sunlight shines through a prism; light separates into its component colours as it travels at different speeds through the prism, a process known as refraction. This optical phenomenon is also seen to occur naturally and is responsible for some of the most unusual optical events connected to the weather.

These effects include the rainbow, of course, but also the famous mirages, coronas and the spectacular light show of the high latitudes, the auroras. The latter, either the aurora borealis in the Northern Hemisphere or its southern relative the aurora Australis, are sometimes better known as the Northern or Southern Lights. Once seen, the rapidly changing coloured lights that illuminate the night sky are impossible to forget.

Perhaps the most famous optical illusion formed by the sun's rays is the mirage. Countless images in cartoon, literature and film highlight this illusory phenomenon.

The less well known corona is formed around the sun or the moon by, once again, light refracting as it passes through the atmosphere.

Left: Fish-eye lens view of the aurora borealis or northern lights display.

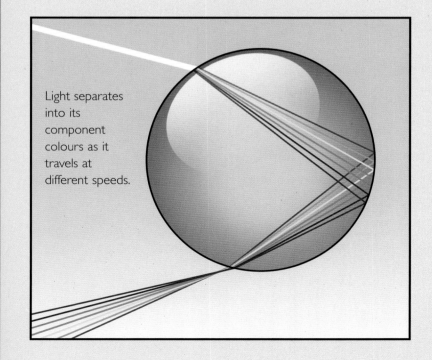

Light separates into its component colours as it travels at different speeds.

Right: Image of a corona around the solar disc. It consists of one or more rings located symmetrically around the Sun and is caused by the diffraction of light passing through water droplets. The rings usually extend for a few solar radii and the width of each ring is inversely related to the diameter of the water droplets causing it. The same phenomenon may be seen around the lunar disc.

Opposite above: Lunar corona, seen through cirriform clouds. A corona is the diffraction of light as it passes through water droplets in clouds. The cirriform clouds are possibly thin high-level cirrostratus nebulosis clouds. The size and regularity of the water droplets dictates the size and colour of the corona.

Opposite below: Rainbows occur when the observer is facing falling rain but with the Sun behind them. White light is reflected inside the raindrops and split into its component colours by refraction. The second rainbow is caused by different raindrops, which are higher in the atmosphere. These raindrops reflect the sunlight twice, causing the order of the second rainbow's colours to reverse.

Rainbows and Coronas

Almost every child has heard the story of the crock of gold at the end of the rainbow but how many know how these features are formed?

A rainbow is formed by the way the light bends as it passes through the atmosphere. A rainbow is sunlight spread out into its spectrum of colours and refracted by water droplets in the Earth's atmosphere. As it is seen as an arc in the sky it lends its name to the bow part of rainbow. Some of the light that has entered the water droplet is reflected allowing the viewer to see the different colours but only if the viewer is standing with his back to the sun.

The traditional view of a rainbow is that it is made up of seven colours – red, orange, yellow, green, blue, indigo and violet – but in fact it is a continuum of colours from red to violet. The colours stem from the fact that sunlight is made up of colours that the

eye can see. When combined, these colours appear white, a principle first demonstrated by Newton in 1666. As light moves through air of different densities it refracts. This disperses the light into its different component colours.

It is when looking towards the sun or moon, however, that a corona can be seen. A corona is a feature which is produced when light is diffracted by small particles, the light passes around the tiny spherical droplets. This is a pattern of lightly coloured rainbow-like bands that surround the sun or the moon when viewed through altocumulus clouds. It's best seen when around the moon as it is less bright than the sun (and safer to look at) so doesn't swamp the effect.

Sometimes a phenomenon known as iridescence appears, irregular patches of colour in mid-level clouds near the sun or moon. The same process is responsible – diffraction of light around water droplets.

Haloes and Sundogs

For many centuries traditional weather forecasters have used the occurrence of haloes as an indicator of imminent rain, but what is a halo?

Airborne ice crystals sometimes act as prisms and when sun – or moon-light – passes through the crystals they refract, that is bend, slightly. As many of the ice crystals may be in random patterns circular haloes result. The most common angle of refraction is 22° and most haloes are produced in this way.

At certain points on the halo, particularly bright patches form. These are known as sundogs, sometimes parhelia or mock suns. Haloes and sundogs occur frequently in polar regions but can also be seen in lower latitudes.

As the refraction of light is also dependent on colour, haloes and sundogs exhibit considerable variation in colouring.

Haloes do sometimes precede rain as the ice crystals responsible for them becoming visible are seen in cirrus clouds, which sometimes precede a frontal system that brings rain.

Below: Rainbow-coloured sundog in a partly cloudy sky in Colorado, USA. This optical atmospheric effect, also called a parhelion, is caused by the refraction, or bending, of sunlight by ice crystals in the atmosphere. It usually causes bright areas on either side of the sun, sometimes referred to as mock suns. Here, the sunlight has been split into the component colours of the spectrum, in the same way that a rainbow forms.

Above: A sun halo is an atmospheric condition that occurs when sunlight is refracted by ice crystals high up in the air. The light is most commonly refracted at an angle of 22°, creating a ring that is 44° in diameter. The shape of the ice crystals also causes two spots of light to appear either side of the halo.

Right: A moondog photographed at Queen Charlotte Islands, Canada. Also known as a mock moon, this optical atmospheric effect is seen as bright areas on either side of the Moon. It is caused by the refraction (bending) of light by hexagonal ice crystals falling through the atmosphere. If the ice crystals are contained within cirrus clouds it may indicate the approach of a frontal system.

Right: Aurora borealis or northern lights display over silhouetted conifer trees in Finland in early October. In the northern hemisphere the aurora borealis is named after the Roman goddess of the dawn, Aurora, and the Greek name for the north wind, Boreas. The aurora borealis most often occurs from September to October, and from March to April.

Opposite above: Aurorae are caused by the interaction between energetic charged particles from the Sun and gas molecules in the upper atmosphere about 100 kilometres/62 miles up. The lights can be white, or coloured red, green, blue and yellow and make many different patterns like rays from a searchlight, twisting flames, shooting streamers or rippling curtains.

Opposite below: Aurora borealis or northern lights display over silhouetted conifer trees in Churchill, Manitoba, Canada. The constellation Gemini is seen at upper right. The colours of an aurora are highly distinctive. The most common colours are green, which indicates that oxygen atoms are involved, and red, which indicate the presence of nitrogen molecules.

Aurorae

Famously seen in the night skies of the higher latitudes is the aurora borealis, or its cousin in the southern hemisphere the aurora australis. They appear as rapidly moving differently coloured areas of light in the night sky. Aurorae occur when particles from the sun interact with the Earth's atmosphere. The particles are called a 'solar wind', which is linked to an 11 year solar sunspot cycle, and solar flares. These are quickly moving charged particles emitted from the sun. The polar regions are the only ones not protected from these particles by the Earth's magnetic field and so it is only at the poles that the solar wind can interact with the Earth's atmosphere.

When the charged particles collide with the Earth's air molecules energy is emitted as light. This is seen only at or close to the poles in a ring called an 'aurora oval'. When there are more particles, such as during a great deal of activity when there are many solar flares, the ring expands and the lights are seen further south or north. They occur very high in the atmosphere – 64 kilometres/40 miles, above aeroplanes – but can occur up to 966 kilometres/600 miles above the Earth which is higher than Space Shuttles fly.

Mirages

How many films of a particular genre made during the 20th century have shown someone in the desert struggling with his thirst seeing an image of a source of water appear before him, only for it to turn out to be a mirage on closer inspection. This idea of a mirage appearing only when someone struggles with heat stroke is erroneous. A mirage is not a mere illusion – it can be photographed – but it is unlikely to be as near nor as tangible as the viewer may hope.

When light passes through different mediums its speed changes, which leads to the light bending or refracting. This can happen in air of different densities such as when the temperature changes.

There are two main types of mirage, called inferior or superior, both of which are caused by the refraction of light passing through the atmosphere. An inferior mirage is seen below an object, such as the sky or the sun, whereas a superior mirage is seen above it.

A commonly seen example of an inferior mirage is the shimmering pool of water often seen on roads on a hot day. The refraction of the light is caused by the very large temperature differences which develop close to the road surface. The light bends so much that an image of the sky shows on the road.

A less common superior mirage often shows itself in the form of the setting or rising sun. It is in fact a mirage of the actual sun which is in reality below the horizon. It is caused by a layer of warmer air above a colder one – a temperature inversion. If the temperature difference is great enough the warmer air refracts the light. Another commonly reported mirage of this type is the ship that appears to be sailing in the sky.

In the USA in 1977 a famous example of a widely seen mirage occurred at Grand Haven where lights were seen across Lake Michigan. The nearest town was Milwaukee, some 120 kilometres/75 miles away but a temperature inversion appears to have created a mirage of the town's lights and make them visible across the water.

Some people believe that superior mirages may be responsible for many UFO sightings – in reality a superior mirage of distant car headlights.

Below: Image of a superior mirage known as a Fata Morgana, created by an island. A double image of the island is seen above it. This type of mirage transforms a fairly uniform horizon into one of vertical walls and columns with spires. According to legend, Fata Morgana (Italian for 'fairy Morgan') was the half sister of King Arthur. Morgan, who was said to live in a crystal palace beneath the water, could build fantastic castles out of thin air across the straits of Messina. Mirages are optical phenomenon occurring when layers of air with different temperatures and refractivities lie close to the ground or water, and are seen when these layers are not mixed.

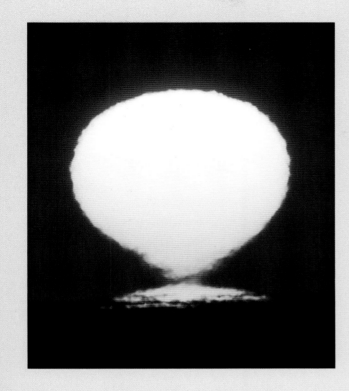

Left: A distorted solar disc at sunset with a mirage forming below it. A mirage is a naturally occurring phenomenon, in which light rays are bent to produce a displaced image of distant objects in the sky. The word comes from the Latin, mirare, meaning 'to appear'.

Below: Desert Mirage photographed in the Western Desert, Egypt. A large mirage glimmers in the heat of the desert sun, casting reflections of the mountains beyond it. A mirage is an atmospheric phenomenon caused by the differential heating of layers of air. As air warms up, it expands, and this affects its refractive index. Light rays slanting down from a cool layer of air to a hot layer (such as one in contact with hot ground) will be be refracted to a shallower angle. If the angle is sufficiently shallow, the light rays are refracted so that they appear to be reflected upwards. Therefore a mirage gives an image of the sky and any scenery behind it.

WEATHER AND CLIMATE IN HISTORY AND CULTURE

Folklore

Today, meteorologists have an impressive array of technological devices and scientific techniques at their disposal, in order to forecast or predict changing patterns of weather and climate. However, for thousands of years, man relied upon a combination of superstition and observation, and over time ideas that particular gods controlled various aspects of the weather were gradually complemented and superseded by looking at natural indicators such as wind direction, cloud formations, the colour of the sky and certain optical effects, and also the behaviour of animals and plants.

The ability to predict the weather was of chief importance to people such as farmers, sailors, fishermen and shepherds, as well as others who lived off the land and the sea, and whose livelihoods, and even lives, often depended upon weather conditions, but across the globe, and particularly in the temperate, higher latitudes of the hemispheres where conditions are most variable, almost every culture developed its own weather lore, which often found expression in proverbs and rhymes.

Some of these have no scientific grounding and rarely prove to be accurate, however many of those relating to the short term, which are based on observation, still hold true to this day. Longer range forecasts, particularly those that are centred around particular days of the year, also tend to be inaccurate; being based on superstitious beliefs, but there are those that developed from long-term observations of seasonal weather patterns and other natural cycles, which continue to hold some truth.

Two of the most well-known weather lore proverbs – 'Red sky at night, shepherd's delight; red sky in morning, shepherd's warning', and 'Mare's tails and mackerel scales make tall ships take in their sails' – both have a basis in truth, and can be explained scientifically. A red sunset often occurs as the result of light being filtered by an approaching high pressure system from the west, which tends to bring fair weather, whilst a red sunrise in the east often indicates that such a system has passed over, and a low pressure system, which is often accompanied by unsettled weather, may be following behind. The second proverb refers to high altitude cirrus and cirrocumulus clouds, which often suggest the approach of a low pressure system that may herald a change in wind speed and direction, and increased precipitation.

Some other fairly reliable short-range indications that have resulted from observing nature include the closing of pine cones and certain flowers, and the curling of leaves prior to heavy rain. Conversely, human and animal hair may straighten as a result of increased humidity and moisture in the air. The chirping of grasshoppers and crickets can also be used to gain a fairly accurate measure of temperature, becoming faster as the temperature rises, and slower as it falls.

In Australian Aboriginal cultures, the weather is believed to be subject to the same supernatural forces that govern the entire continuum of life and the natural world, and there are numerous observations concerning the movements of animals and birds, and the flowering of particular plants, which are often good indicators of the onset of the wet and dry seasons.

In North America and elsewhere, however, certain predictions relating to animals and the weather are typically less reliable. On 2 February, which is a cross-quarter day, halfway between the winter solstice and the vernal equinox for example, 'Groundhog Day' is traditionally celebrated. According to weather lore, if the groundhog or marmot emerges from his burrow on that day and casts no shadow, then winter is coming to an end, but if the weather is sunny, the animal will be scared of its own shadow, return to its burrow, and winter will endure for six further weeks. It is thought that this custom may have originated in medieval France or Germany, where the spring equinox, which denoted spring's arrival, occurred exactly six weeks after 2 February.

In the UK, William Foggitt, now aged 88, has been using family records from 1830 to produce forecasts. These are based on observations of the past weather's effects on birds, plants and animals and he currently claims accuracy rates of 88 per cent.

Opposite: Altocumulus clouds forming a mackerel sky at sunset. This spectacular pattern may indicate an approaching weather front and impending rain.

Right: Pine cones on a branch of sitka spruce in British Columbia, Canada. Pine cones have traditionally been used to forecast humidity. Place a pine cone outside where it can be observed regularly. As the humidity increases it will close up to protect the seeds.

Ancient Beliefs

Many people will know that Ra is the ancient Egyptian sun god but he is only one of many dozens of others found in different cultures at different historical periods. In an attempt to understand nature and the weather in particular many ancient peoples devised 'weather gods' who controlled the wind, sun and rain, who shaped the harvests and the availability of wild animals for food. Early people were superstitious and fearful but believed that gods would be appeased by offerings and rituals.

It is believed that human sacrifices were made to the Aztec rain and harvest god Tlaloc whereas Quetzalcoatl was believed to be the creator of life and in control of the vital rain-bearing winds. Other cultures had gods of wind such as Aeolus in ancient Greece and Fujin a Japanese Shinto demon. The Japanese World War Two 'Kamikaze' (divine wind) pilots took their name from the name of a god of storms and winds. Zeus (Greek), Raiden (Japan) and Indra (Hindu) were all gods associated with thunder and, particularly, lightning. The vital life-giving sun was worshipped by many cultures including the Egyptians and the Incas: the Egyptian Ra was depicted as travelling across the sky in a sun boat each day, the Incan Inti was personified as a golden disc with a human face while. The Mayan culture worshipped the rain god, Chac. Even today Mayan descendants perform a rain ceremony called Cha-Chac every May towards the end of the dry season.

In many other places across the world today there are still traditional festivals which take place in an attempt to predict what conventional forecasters cannot.

Below: The tomb of Antiochus in Nemrut Dagi, Turkey. At this mountaintop shrine to the gods, the heads are 2.4 metre/8 feet tall representations of ancient Greek gods. In the foreground are Hercules, Apollo and eagle Zeus, the Greek god of thunder. They were built by a pre-Roman king, who cut two huge ledges into the mountain-top and filled them with statues of himself and various gods.

Above: An Aztec calendar of the sun. The picture of the sun in the centre reflects its importance in the life of the Aztec people. Each season is shown on the perimeter of the stone. The Aztec calendar consisted of a 365-day calendar and a 260-day ritual cycle. The two cycles together formed a 52-year 'century', sometimes called the 'Calendar Round'.

Above: Quetzalcoatl was a Toltec and Aztec god. He is often associated with the wind god, Ehecatl, and represents the forces of nature. Quetzalcoatl also became a representation of the rain, the celestial waters and their associated winds.

The Renaissance and Early Science

Greek scholar Aristotle's book *Meteorologica* was perhaps the first scientific attempt to understand the weather. Very early scientific attempts to explain the weather were carried out by Theophrastus, a student of Aristotle, who wrote *On Weather Signs*. Later the work was continued by the Romans; Pliny the Elder wrote *Historia Naturalis* in which he drew together records and superstitions from around the ancient world. Following the collapse of the Roman Empire meteorological research was confined to Islamic scholars. In the 10th-century *The Chronicle of Ancient Nations* was written by al-Buruni and contains a monthly calendar of weather and climate information derived from ancient Greek, Egyptian and Roman sources.

Meteorology did not progress in Europe until the Renaissance period in the 15th century, a time of immense change and developments in understanding culture and science. One area which benefited during this time was the development of new and improved scientific instruments for measuring and understanding the weather. During the 15th century the expansion of trade routes led to a greater understanding of the Earth and the weather. Magellan, Columbus and other explorers of the period began to recognize the pattern of winds across the oceans and kept records of the weather. The Earth had been circumnavigated twice by 1600.

Significant figures of the time include Galileo Galilei (1564–1642), Leonardo daVinci (1452–1519) and Evangelista Torricelli (1608–47). Galileo is perhaps most famous in this context for the invention of a thermometer; he was convinced that many weather phenomena could be explained by scientific means. daVinci pioneered the study of atmospheric optics and was responsible for encouraging an interest in understanding the forces of nature. He was based in Florence, the hub of scientific investigation during the Renaissance; from 1657–1667 the Accademia del Climento developed a number of weather recording instruments. Torricelli invented a barometer known as the 'torricellian tube'. Blaise Pascal, a French scientist, realized that changes in atmospheric pressure could be measured by a barometer and could be related to weather changes. Many of the later developments in understanding and measuring the weather stem from the pioneering work of this period.

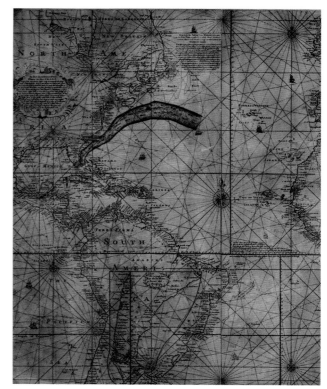

Opposite below left: Leonardo da Vinci was born in 1452, near the town of Vinci, not far from Florence. Despite having little formal education, he kept extensive notes with his scientific theories, inventions and architectural designs.

Opposite below right: Franklin-Folger chart of the Gulf Stream on a map of the Atlantic Ocean published in 1769. Benjamin Franklin asked his cousin Timothy Folger, captain of a whaling ship, to produce this map to help shipping make faster crossings. Navigation instructions are at upper centre and upper left. The map title credits Captain Edmund Halley (he of comet fame) with the main map and Mount and Page as the publishers.

Right: Sir Isaac Newton (1642–1727), astronomer, mathematician and physicist. Newton made a number of profound advances in a range of disciplines. His laws of gravity and motion formed the basis for the model of the universe until Einstein. He invented the reflecting telescope, which is used in nearly all large observatories, and studied the spectrum of light showing it was made of a mixture of colours. He showed that light could be bent by prisms, and devised laws of reflection and refraction. In mathematics, he devised the binomial theorem, and developed differential and integral calculus.

The 17th and 18th Centuries: The Age of Reason

Pascal, Newton, Celsius; all names we associate with different important scientific discoveries during the 17th and 18th centuries and all related to the improvement in measuring and forecasting the weather which began during this time.

The French scientist Blaise Pascal discovered that air pressure changed with height. He also realized that air pressure was related to changes in the weather.

In England, Isaac Newton, most famous for identifying gravity, devised laws of physics that remain the basis of today's weather forecasting computer models.

A German Physicist, Gabriel Fahrenheit developed the eponymous temperature scale in 1724. Fahrenheit dedicated much of his time to developing meteorological instruments, including reliable alcohol and mercury-based thermometers, which he produced commercially from around 1717 until 1736, the year of his death. There are conflicting accounts of exactly how his scale was devised, but it is thought that on a scale using divisions of twelve, and later subdivisions of eight, he took zero degrees to be the point registered by placing his thermometer in a bath of ice, salt and water, 32° to be the point at which water froze, and 96° to be the temperature of the human body, measured in the mouth or armpit. Some however, contest that he adopted a previous scale developed by the Danish astronomer Ole Rømer, and simply multiplied the existing values to remove the fractions, before calibrating his scale according to the methods previously described. Yet regardless of quite how the Fahrenheit scale was arrived at, it proved to be a success, remaining in common use for meteorological, medical and industrial applications until the widespread adoption of the Celsius scale during the 1960s and 1970s. Some continue to maintain that the Fahrenheit scale is more easily understood, as its measurements can be easily conveyed without the use of fractions or decimals, and that these relate more closely to general perceived air temperatures. The Fahrenheit scale continues to be used, particularly for meteorological purposes, in the US.

The Celsius scale was originally created in 1742 by the Swedish astronomer Anders Celsius, taking zero as the boiling point of water, which, he accurately demonstrated, varied according to atmospheric pressure, and 100 degrees as the melting point of ice, which did not. The boiling point of water was then calibrated at the average barometric pressure at sea level. This scale was effectively the first attempt to introduce a standardized, international scale, which did not require the expression of negative or minus values in recording extremely cold temperatures. Two years later in 1744, the same year that Celsius died, the scientist Carolus Linnaeus is thought to have then reversed Celsius's scale, so that zero degrees represented the melting point of ice, and 100 degrees denoted the boiling point of water. However, several individuals have been credited with this achievement, including Celsius's student and successor Mårten Strömer, and the Swedish instrument maker Daniel Ekström, who is known to have supplied thermometers to Linnaeus. Today, it is the reversed Celsius scale that persists, and which is used throughout the scientific community worldwide. It has also become generally accepted for non-scientific purposes, having been introduced as part of the international metrication process of the 1960s and 1970s. Since 1954 however, the Celsius scale has been related to the Kelvin scale within the scientific world, with 'absolute zero' and

Left: Portrait of the Swedish astronomer Anders Celsius (1701–1744). Celsius was the first to associate the aurora borealis with changes in the earth's magnetic field. In 1730 he became professor of astronomy at Uppsala, and by 1740 he was in charge of the large new observatory there. However, his best known achievement was his temperature scale, which divided the temperature difference between the boiling point and freezing point of water into an even 100 degrees. He first described this in 1742 when he placed the boiling point at zero and the freezing point at 100 degrees. The following year this was reversed. This scale is used by scientists everywhere.

Opposite above: Illustration of Benjamin Franklin and his son William, performing their famous experiment on 15 June 1752, flying a kite in a thunderstorm. A metal wire on the kite attracted a lightning strike and electricity flowed down the string to a key, charging a Leyden jar (capacitor) seen near Franklin's hand. This experiment proved that lightning was an electrical phenomenon, and supported Franklin's invention of lightning rods. Some scientists died repeating the experiment, and the Franklins were lucky they were not killed themselves.

the 'triple point' of scientifically standardized water replacing the melting point of ice and the boiling point of water.

During this period the wealthy and educated were able to take advantage of the newly developing weather recording instruments to make their own observations. Thomas Jefferson, the third president of the USA took weather readings from 1776 to 1816. If legend is to be believed, he acquired his first thermometer while writing the Declaration of Independence. Another pioneering American philosopher-statesman, Benjamin Franklin, went about meteorological experimentation in an altogether different manner. In 1752 he conducted a landmark – and highly dangerous – experiment to see if lightning was in fact an electrical discharge. He flew a kite in an electrical storm and was lucky to survive. He is also believed to be the inventor of the lightning rod which protects buildings from direct strikes.

Temperature Conversion Chart

Celsius °C	Fahrenheit °F
0	32.0
1	33.8
2	35.6
3	37.4
4	39.2
5	41.0
6	42.8
7	44.6
8	46.4
9	48.2
10	50.0
11	51.8
12	53.6
13	55.4
14	57.2
15	59.0
16	60.8
17	62.6
18	64.4
19	66.2
20	68.0
21	69.8
22	71.6
23	73.4
24	75.2
25	77.0
26	78.8
27	80.6
28	82.4
29	84.2
30	86.0

The equation for converting Fahrenheit to Celsius is:

$$((\text{Deg. F}) - 32) \times (5/9) = \text{Deg. C}$$

Far left: William Kelvin who proposed the absolute or Kelvin temperature scale in 1848.

Left: The Fahrenheit scale was based on 0° as the point registered in a bath of ice, salt and water, 32° as the point at which water freezes, and 96° as the temperature of the human body. In the Celcius scale, 0° is the melting point of ice and 100° the boiling point of water.

The 19th Century:
The Mariners' Influence

The massive expansion in colonization and trade which occurred during the 18th and 19th centuries is, arguably, responsible for the development of and the practical use of early weather recording instruments and the increase in understanding of the weather.

In 1806 the British Admiral Francis Beaufort developed the Beaufort scale for measuring wind speed. Although not everyone at the time appreciated its use, it was adopted by the Royal Navy in 1838 and time has shown us how valuable it still is today.

A little later than Beaufort, Royal Naval Captain (later Admiral) Robert FitzRoy was responsible for developing a type of barometer that could be mass-produced and which was distributed to every port around the British Isles. He had observed that weather systems tended to move from west to east and that low pressure could be tracked by use of an Atlantic-wide system of weather stations. The invention of the telegraph system during this time made the dissemination of the information much easier. Once appointed as Head of Meteorology, FitzRoy was able to provide the first national weather forecasts at the same time as Captain Matthew Maury was creating a similar system in the USA – as were others in France. FitzRoy's forecasts were published in *The Times* in London from 1860 and Queen Victoria requested a personal forecast prior to a visit to her holiday home on the Isle of Wight. Despite some setbacks FitzRoy's system was eventually approved and formed the basis of modern weather forecasting. His barometer is still in production today.

Right: Illustration of a weather station at the Greenwich Royal Observatory, London in 1880. The station contained an automatic anemometer, which measured the direction and strength of the wind, and recorded it on a strip of paper. The science of meteorology advanced greatly during the late 19th century due to the development of telegraphy. This allowed data to be quickly collected from many weather stations at once. The first daily weather maps were published in 1860.

Opposite: Meteorological archive showing 19th-century daily meteorological record from Woolwich, UK. On the left-hand page is a key of the abbreviations. The Meteorological Office in Exeter, Devon, holds reports for the UK for every day from 1 January 1869 to the present.

Left: Portrait of Robert FitzRoy, RN (1805–1865), British hydrographer and meteorologist. FitzRoy was educated at the Royal Naval College at Portsmouth. He commanded HMS *Beagle* during both of its survey voyages to South America (1828 and 1831). During the second of these, FitzRoy took along Charles Darwin as the ship's naturalist. His narrative of the voyages included *The Voyage of the Beagle*, written by Darwin. FitzRoy served as an MP from 1841–43, then as Governor of New Zealand until 1845. In 1854 he took up the post of meteorological statist, organizing a number of observation posts and the publication of storm warnings. He committed suicide in 1865.

The 20th Century

The early 20th century marked the start of a transformation in the methods available to forecasters. One of the earliest pioneers in using mathematical techniques for forecasting was Lewis Fry Richardson. His interest in meteorology led him to suggest a scheme for weather forecasting using differential equations. This is the method used today, but via powerful and fast computers not available to Fry at the time. When he wrote *Weather Prediction by Numerical Process* in 1922 he envisaged speeding up the calculation processes by using messengers on motorbikes to transfer information from one stage of the calculation process to another. He was also interested in atmospheric turbulence and the Richardson number, a dimensionless parameter in the theory of turbulence is named after him. Many students of meteorology or physics will recognize his famous rhyme:

Big whorls have little whorls that feed on their velocity,

And little whorls have smaller whorls and so on to viscosity.

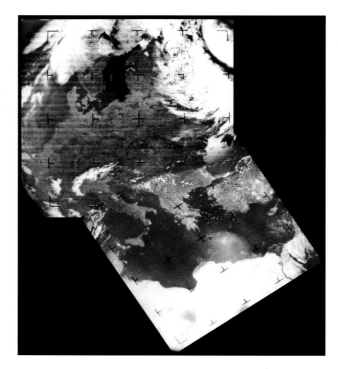

Above: Early weather satellite images of Europe from the TIROS (Television and Infrared Observation Satellite) programme (1960–1966), which used television cameras to monitor cloud patterns and the Nimbus satellites (1964–1978). These satellites were the first in a long succession of various weather satellites. Modern equipment is able to gather a much wider range of data than these early weather satellites did.

The early years of the century also saw the beginnings of electronic computers, the modern versions of which we rely upon today for our forecasts. In the 1930s aircraft were used to take upper atmosphere observations, a significant meteorological breakthrough. Because upper atmosphere pressure, wind and temperature have a strong impact on ground level weather patterns, this was a major advance in forecasting. Another means of measuring upper atmosphere conditions is radiosonde or weather balloons, first developed during the same period but expanded rapidly during the 1940s. Radar began as a by-product of World War II technology to track enemy aircraft during the Battle of Britain in 1940. This is now used widely to track the oncoming rain showers and is most widely used in the USA to track tornadoes and hurricanes.

Opposite: Scientists working on the Tiros satellite, the first weather satellite. Tiros (Television Infrared Observation Satellite) was launched on 1 April 1960. Its images of clouds led to a greater understanding of weather systems.

Above: Since the beginning of the 'space age' the technology used in weather forecasting has advanced significantly. Here technician tests the sounder, part of the GOES-L weather satellite.

Historical Events Influenced by the Weather

We all have had days out ruined by unexpectedly poor weather, but for most of us these are not life threatening nor of national or international significance. However, the course of history has been altered by the vagaries of the weather.

Julius Caesar invaded Britain in 54 BC, but this was not his first attempt to conquer the the British Isles. In 55 BC his navy was thwarted by the strong northwesterly winds, which made crossing the English Channel impossible.

A later invasion of Britain, this time by the Spanish Armada in the 16th century, was also affected by the weather. In the late summer of 1588 the Spanish ships retreated after defeat by Francis Drake's naval forces. A deep depression bringing strong winds and very high seas in the north of Scotland and Ireland wrecked the remaining ships, drowning many hundreds and effectively ending Spanish ambitions to mount a further attempt on Great Britain.

Throughout history, there are numerous similar examples of the weather's effect on the outcome of military campaigns. For example, during the 13th century the naval forces of the Mongol ruler Kublai Khan were twice thwarted in their attempts to invade Japan, in 1274 and 1281, with the fleets being ravaged by typhoons on both occasions. These winds became known in Japanese history as the 'kamikaze' or 'divine winds'.

During the mid 17th century in northern Europe, when Denmark attacked Sweden, severe winter weather enabled the Swedish forces to cross the frozen ice from Jutland to Zealand, forcing the Danes to surrender, and in 1812, Napoleon's Grand Army was forced to retreat from Russia, partly on account of the severe winter weather. Only two years later, at the Battle of Waterloo, thunderstorms, accompanied by torrential rains, forced Napoleon to delay his attack on the British, which enabled Prussian forces to join them, whilst the French advance was slowed considerably, and the battlefield turned into a quagmire. It is often suggested that Napoleon may have been victorious had he been able to launch his attack earlier.

More recently, during World War II, the weather played a significant part on at least two

occasions. Firstly, during the winters of 1941 and 1942, the German forces were hampered in their assault on Russia, and the Soviet army was able to launch its first decisive counter-offensives at Moscow and Stalingrad.

Then later, the D-Day landings of June 1944 saw the Allied governments relying very heavily on forecasters' predictions of the coming weather. The commanders needed a 'weather window' to land troops safely, to fly in gliders and for bombers to be able to provide air cover. Each needed slightly different weather conditions so it was difficult to find a suitable date for all concerned. Following an unsettled period of weather, forecasters accurately predicted that it would clear up for long enough on the night of 5 and into 6 June. The forecasters were right, the decision to invade was taken and the rest is, well, history.

Other historical events in which the weather has played a key part include the Great Fire of London in 1666, when fire swept through the city following a prolonged dry spell. The tinder-dry wood and thatch buildings were easily set alight and the flames were fanned by a strong easterly wind making the fire spread rapidly. Amongst the buildings lost was the original St Paul's Cathedral. Drought conditions during the spring and summer of 1788 in France, which led to a failure of the harvest and widespread famine, followed by a severe winter, contributed significantly to the peasant uprisings that were to culminate in the French Revolution during the summer of 1789.

Opposite: A lithograph depicting Napoleon's return from Moscow, after an original painting by Jan Chelminski. The Russian winter combined with disease, desertions, and casualties sustained in various minor actions caused thousands of losses in the French camp.

Above right: An oil painting of Old St Paul's Cathedral (1616) before it was destroyed in the Great Fire of London in September 1666.

Right: Soldiers marching through the ever deepening snows in single file in Russia in December 1942.

Weather in Literature and Art

From ancient scriptures to modern novels, and in both prose and poetry, the weather has been a recurrent subject in a great deal of literature throughout the ages. It is frequently employed in fiction, not only observationally or as a simple aesthetic aside, but as a device for setting tone and emotional atmosphere, and it may not only inform the plot of a story, but actually drive and parallel it. Meanwhile, in the world of science fiction, certain ideas about harnessing and controlling the weather have proven strangely prophetic.

The god of both war and weather, Indra, features prominently in the ancient Indian Vedas – collections of sacred Hindu writings and knowledge, which are thought to have been first written down around 3000 years ago. The Bible is a rich source of weather-related references, in both the Old and New Testaments. Perhaps one that most people will be aware of is the story of Noah and the great flood as written in Genesis Ch 6–9. which is also paralleled in the Babylonian 'Epic of Gilgamesh'. The flood is described as a deliberate act by God to cleanse the Earth of life because 'the earth was corrupted before

God, and was filled with iniquity'. Noah, his family and the animals he collected were saved by building an ark or boat in which to survive the flood, which lasted 40 days and 40 nights. The weather was awful: 'all the fountains of the great deep were broken up and the flood gates of heaven were open'. Whether such a flood actually occurred is open to conjecture: many modern climatologists believe the events depicted in Genesis (and in other mythologies from other cultures) reflect a sudden and major melting of ice – a period of global warming.

In marked contrast are Biblical references to drought such as that from, once again, Genesis relating to the dreams of the Egyptian Pharaoh. Joseph, a young Jew, was sold to the Pharaoh, who referred to Joseph when troubled by dreams of impending feast followed by famine in Egypt. Given the precarious climate of the region it is likely that the cause of the famine was a prolonged drought. That Joseph made plans to prepare for the famine reflects that society's familiarity with the problem.

The book of Exodus talks of the escape of the Israelites, led by Moses, from the slavery imposed on them by the Egyptians. To prove to the Pharaoh that God existed and to persuade him to release the Israelites from slavery, Moses asked God to perform a series of miracles. One of these was a hail storm before which Moses warned the Pharaoh that God would 'cause it to rain a very grievous hail'. When this happened it caused a great deal of damage to homes, livestock and crops. The likely cost of a similar storm in Biblical times must have been very great.

Weather is also an important feature of the Norse Eddas of the 13th century; the god Thor was amongst the most powerful of the gods, and according to legend, he rode through the heavens in a goat-drawn chariot during thunderstorms, producing thunder and lightning by striking and throwing his war hammer, Mjollnir. However, there was also a distinct god of the weather, Vidrer.

In the dramatic works of William Shakespeare, written in the late 16th and early 17th centuries, the weather is frequently an integral element, and it was often employed symbolically. For example, he often used stormy weather as a metaphor for the difficulties of human relationships, or to parallel violent events, as in the tragedies *Macbeth*, *Julius Caesar*, and *King Lear*. Storms are also used as plot devices, as in *Twelfth Night*, where a storm forces the separation of the characters Viola and Sebastian, and conversely in *The Tempest*, where a storm brings the characters together. In this play, the power of nature and the weather over humans can also be seen as a major theme.

During the 18th century, the German poet, novelist, dramatist, philosopher, theorist and scientist Johann Wolfgang von Goethe was influential in many fields. He also brought the work of Luke Howard, who was the first to classify different types of clouds, to the attentions of numerous artists, who, along with the Romantic poets such as William Blake, Lord Byron, Samuel Taylor Coleridge, Percy B. Shelley, William Wordsworth and John Keats, formed the nucleus of the Romantic movement in England, which had counterparts throughout much of Europe. Weather and the changing seasons were a primary

Above: William Shakespeare, who often used the weather as an integral part of the action in his plays. Storms are used as plot devices, to parallel violent events or to signify difficult human relationships.

Left: A portrait of the German writer Johann Wolfgang von Goethe.

Opposite: A scene from *The Tempest*. Shakespeare's play opens with Prospero, rightful Duke of Milan, raising a storm in order to run his brother Alonso's ship aground.

feature in many of the works of these poets, as revealed by the titles of works such as 'The Cloud' and 'Ode to the West Wind', by Shelley, 'I Wandered Lonely as a Cloud', by Wordsworth, and 'To Autumn', by Keats.

In St Petersburg, Russia, a similar tradition can be found in the literary works of Nikolai Gogol, Fyodor Dostoevsky and Nikolai Alekseevich Nekrasov, during the mid 19th century. The typically bad and unstable weather described, which was a real feature of the city, often reflects the inner emotional state of the characters or a wider social malevolence, and it may also drive the plot, overwhelming characters, imbuing them with melancholy and bringing about their misfortunes – as in Nekrasov's 'About the Weather', for example. Similarly, the late 19th century works of the English novelist Thomas Hardy are thoroughly imbued with often bleak and unforgiving weather, which not only provides a dramatic backdrop, but again often determines the fate of the characters.

The weather, and in particular the idea of controlling or creating it, is a common feature of many science fiction novels, and the concept of terraforming, or modifying the atmosphere and ecology of a planet, including its weather, in order for it to become habitable by man, is a recurrent theme. Such ideas remain theoretical at present, but similar ones, such as many of those proposed in the American inventor and writer John Jacob Astor IV's *A Journey in Other Worlds*, published in 1894, have startling parallels in the real world today. For example, he writes about generating electrical power by means of large-scale wind farms and structures equivalent to solar panels, and also describes personal windmills on the roofs of houses, which are connected to storage batteries to provide energy for heating and lighting.

Art and the Weather

The weather and its elements have featured in artistic representations for thousands of years; even being known from prehistoric pictographs and petroglyphs, which were painted and carved in caves and on other rock faces. From these earliest depictions, there subsequently developed a long tradition of representing the natural world and its weather, particularly in sky and landscape painting, which continues to this day. However, today there are also artists who incorporate natural weather phenomena into their art, replicate them artificially, or use meteorological data to produce their works.

Above: Artwork of a family being rescued from
the roof of their house after a river flooded the
surrounding area. The image is set in the South of
France and was drawn in 1875.

Opposite: 'Hunters in the Snow' (1565) by Pieter the
Elder Brueghel. Brueghel was a Flemish Renaissance
painter known for his landscapes and peasant scenes.
He is often credited as being the first western painter
to paint landscapes for their own sake, rather than as a
backdrop to a religious allegory. His winter landscapes
of 1565 are corroborative evidence of the severity
of winters during the 'Little Ice Age'.

Right: Mughal miniature painting depicting a lady
brought in from a night storm dating from
around 1760.

It is thought that realistic sky painting began with the ancient Greeks and was continued by the Romans, but it also features prominently in ancient Oriental paintings. In Japan this tradition was continued in the 18th and 19th centuries by such artists as Matsumura Goshun, Okamoto Toyohiko, who painted 'Seashore and Mountains in the Seasons', Katsushika Hokusai, responsible for, amongst many weather-inspired works, 'Mount Fuji Above the Lightning' (1829–33), and Ando Hiroshige, whose 'Sudden Shower at Ohashi Bridge at Atake' (1857), is thought to have informed Vincent Van Gogh's 'Rain at Auvers' (1890).

The first major explosion of weather in art, however, resulted from the increased interest in observing nature that was introduced during the Renaissance of the 15th and 16th centuries, with artists such as Leonardo da Vinci, Rembrandt, Titian, Albrecht Dürer, Jan van Eyck and the Brueghels, all noted for their rendering of clouds and skies.

Later, during the 18th and 19th centuries, the weather became a major concern of the Romantic artists in western Europe, many of whom produced symbolic, but naturalistic depictions of the landscape. The painter Caspar David Friedrich, for example, is thought to have been highly influenced by Luke Howard, who invented a classification for clouds in the early 19th century, painting numerous scenes in which snow, the sky and clouds are significant features, such as in 'The Wanderer Above the Sea of Fog' (1818). Other notable exponents of Romanticism include Joseph M.W. Turner who exhibited 'Rain, Steam and Speed' in 1844, and John Constable, who painted numerous studies of clouds during the 1820s.

Later, the effects and changing qualities of sunlight became of prime concern to the 19th-century French Impressionists, who also adopted the technique of painting outdoors, or *en plein air*, amongst the elements. Claude Monet and Camille Pissarro are amongst those artists that perhaps best exemplify the Impressionist movement and works such as Pissarro's 'The Road Under the Effect of Snow' (1879), and Monet's series of haystacks of the early 1890s demonstrate a fascination with changing light and weather conditions.

More recently in the 1970s, the land artist Walter de Maria created 'The Lightning Field' on a plain in New Mexico. It consists of 400 polished stainless steel poles, fixed vertically in the ground in a grid measuring approximately 1 mile/1.6 kilometres

Above: 'Drum Bridge and Setting-Sun Hill at Meguro' by the Japanese artist Ando Hiroshige. Hiroshige was one of the last great artists in the ukiyo-e tradition: a genre of Japanese woodblock prints and paintings produced between the 17th and the 20th century, many of which featured the landscape and the weather.

Opposite: 'Waterloo Bridge, Morning Fog' by Claude Monet. Many of Monet's most famous paintings show the effect of changing light and weather on the landscape.

by 1 kilometre/0.6 mile, with each pole being 67 metres/220 feet apart. Some of the most iconic images of the piece, which were captured by the work's exclusive photographer, John Cliett, are those of lightning striking close to the poles or the poles themselves, but the work is essentially designed to be experienced in situ in all weathers, and is brought to life not only by lightning, but equally by the effects of sunlight, mist and fog.

Similarly, the 'weather artist' Ned Kahn has produced several works that are animated by the weather, whilst simultaneously making its elements, particularly the wind, visible to the onlooker. Such projects include 'Slice of Wind' (1996), 'Wind Veil' (2000), 'Articulated Cloud' (2004), and 'Wind Leaves' (2006), in which thousands of small metal discs or panels, which may be attached to sculptures, or take the form of shade-screens on building facades, are free to rotate in the wind, often dramatically also reflecting the sky and sunlight as they do so. Other works such as 'Prism Tunnel' focus more exclusively on refracting light. However, perhaps Khan's most remarkable works are those in which he replicates natural weather phenomenon by artificial means; producing clouds and fog, and even tornados with fans and water vapour, such as the seven-storey tornado he created indoors in the Duales Systems Pavilion at EXPO 2000, in Hannover, Germany, in 2000.

Olafur Eliasson also creates works which simulate elements of the weather, such as a rainbow, in his work 'Beauty' (1994), fog, as part of the installation 'The Mediated Motion' (2001), and a sunrise and mist in 'The Weather Project' (2003). However the artist also incorporates natural phenomena such as sunlight in his works, as in 'Your Sun Machine' (1997).

Andrea Polli is a sound artist, who works in digital media in conjunction with meteorologists and other scientists, in order to produce sonifications and visualisations, which are based on actual weather data from around the world, and also that which is produced by detailed computer models or simulations of projected weather and climate. Her works include 'Atmospherics/ Weather Works', 'Heat and the Heartbeat of the City' and 'N'.

MODERN WEATHER FORECASTING

Modern Methods of Forecasting

Modern weather forecasting uses a mixture of tried and tested weather recording instruments to provide reliable data which is then supplemented with data from satellites and radar. This data may come from automated stations or from a very large number of local manual weather stations which feed data into a national picture. Once the data is received at the various meteorological offices around the world it is processed largely by computer and inputted into one of a number of computer models that have been developed as an aid to forecasting. Increasingly forecasters are relying on ever more complex models for their predictions. The data they use may come from numerous places around the world; the various meteorological offices share information in almost real time and this reflects the unique and high degree of cooperation that exists in this field of science, originating from the days of Robert FitzRoy and his American and French colleagues in the mid-19th century.

All weather stations, automated or manual, contain standard recording equipment. Inside the Stevenson's Screen, a white, louvred box on legs, are various thermometers, a barometer for recording air pressure and a hygrometer to measure humidity. In addition to these, there will be a rain and/or snow gauge, an anemometer to measure wind speed, a vane to measure wind direction and a sunshine recorder. Each will be read in a strict routine and the data sent electronically to a central meteorological office. From there each set of data is used to predict the weather and may be used as part of the data required for computer models to make longer-term predictions.

In addition to these readings, weather data comes from a variety of other sources. Helium-filled weather balloons provide upper atmosphere data on broadly the same lines as ground-level observations from weather stations. Their results are transmitted to ground stations by radio hence their alternative name: radiosonde balloons. Specially equipped aircraft also collect data on temperature and humidity and can be used to study the physics of clouds. Weather ships carry the same sort of recording instruments as weather buoys and feed information, often essential, to the forecasters. The buoys are sometimes moored or may drift with the ocean currents and are sometimes solar powered. These provide vital information for the forecasters, particularly for maritime countries.

Weather Recording Instruments

Many homes have a barometer or a thermometer, which are still used by modern meteorologists. However, many of the other scientific instruments used for forecasting are rather more sophisticated. The early instruments were often beautifully made in high-quality hardwoods and brass: these are now highly valuable as collectors' items. Without the modern version of these instruments, however, the forecasts we rely upon every day would not be possible. Many of them have changed little since they were first invented.

Thermometer
Modern thermometers use either alcohol or mercury to measure temperature. These liquids are used because they expand when heated and contract when cooled. They both do this at a regular, measurable rate which makes them ideal for this use.

Stevenson's Screen
A number of weather-recording instruments are kept in this specially designed, louvred box where they are protected from the interference of direct sunshine and rainfall.

Barometer
This instrument is used to measure air pressure which is usually shown graphically on a barograph. Air pressure is measured in millibars.

Above: A scientist collects the water from a rainfall sampler to be sent for analysis. Measurement of the relative amounts of the oxygen-18 and hydrogen-2 (deuterium) atomic isotopes in rainwater helps to determine the global circulation of water.

Opposite: Meteorologist launching a weather balloon. The balloon, which is filled with helium, carries aloft an ozonesonde, an instrument that detects ozone in the atmosphere and sends its measurements back to a base station using a radio transmitter. As the balloon gains altitude, and the local pressure drops, the balloon expands. and eventually bursts. The ozonesonde is returned to Earth by a small parachute. The measurements will be used to study the depletion of the polar ozone layer.

Left: Mobile weather radar being used to monitor the approaching gust front of a dust storm. Radar measurements are used to advise people in the path of the storm of their best course of action.

Rain Gauge
A specially calibrated container measures the amount of precipitation that falls in an area.

Hygrometer
This is used to measure humidity in the atmosphere.

Sunshine Recorder
Used to measure how much uninterrupted sunshine occurs in an area.

Anemometer
From the Greek word anemos, or wind, an anemometer measures wind speed on a number of different scales: kph, kmph, mps, mph, or on the Beaufort Scale.

Radiosonde Balloon
Upper atmosphere measurements of temperature, pressure and humidity are needed for forecasters to gain a full set of data from which to produce a forecast. These measurements are taken by miniature instruments placed inside a lightweight box taken into the upper atmosphere by a hydrogen-filled balloon. This is tracked by radar and data is sent by radio signals back to a central computer where it is processed.

Measuring Temperature – Thermometers

There are several different types of thermometers in weather stations that are used by forecasters.

The standard thermometer contains either mercury or alcohol: both liquids react with temperature and expand or contract at a regular, recordable rate allowing accurate measurement of temperature to be taken.

A maximum–minimum (max–min) thermometer contains two separate thermometers side by side. Each one measures the current air temperature as well as the maximum – or minimum – recorded over a particular period. The small metal markers contained inside each thermometer are pushed by the liquid and remain at the maximum or minimum temperatures reached over a chosen period of time, usually one day. The markers can be reset once the temperature has been read, by either a magnet or a push-button mechanism. This type of thermometer is sometimes known as 'Six's thermometer'.

Another type of max–min thermometer is the bimetallic dial thermometer. This has pointers on a dial which are very easy to read. It also needs to be reset after it has been read.

Occasionally a thermograph is used, which shows temperature as a continuous line on a specially designed sheet of graph paper. Each sheet records a single day's temperatures.

The dry bulb thermometer is mercury-filled and records changes in temperature in degrees and tenths of a degree centigrade. This is the temperature shown on television and radio forecasts. The wet bulb thermometer is an ordinary one with a muslin wick covering the bulb. This is fed by water from a special attached reservoir. The comparison between the dry and wet bulb temperatures allows the dew point and so the relative humidity to be calculated

There are, of course, several other types of thermometer and one of those is the Galileo thermometer, an attractive object which is used as a decorative as well as practical item in homes. It is a sealed glass cylinder containing a clear liquid in which are suspended a number of weights. The weights themselves often contain coloured liquid which gives them an attractive appearance. As the liquid changes temperature it changes density and the suspended weights rise and fall. They stay at a position where their density is equal to that of the surrounding liquid. If each weight differs by a small and measurable amount and they are ordered so that the least dense is at the top and the densest at the bottom it is possible to form a temperature scale. Usually each weight has an engraved metal disc attached from which the temperature can be read.

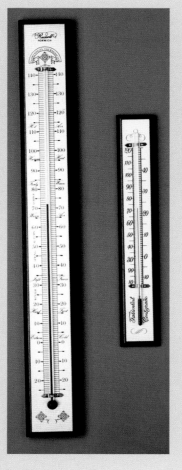

Left: Two thermometers for measuring air temperatures. They each have a hollow, evacuated glass column with a bulb at the bottom, which contains mercury that is dyed blue or red. The mercury expands as the air temperature increases. This means that the level of the mercury rises up the tube and can be read off using a scale.

Opposite: Readings are taken every three hours from this row of Stevenson's screens at the Arctic Research Station in Finnish Lapland.

Left: A maximum and minimum thermometer is used to measure the minimum and maximum air temperature. It has two columns connected by a 'U'-bend at the bottom, which contains mercury. The left column (minimum) leads to an alcohol reservoir. As the temperature increases, the alcohol expands, pushing the left side of the mercury down and forcing the right side up. As the temperature decreases, the alcohol contracts, allowing the mercury to return. Small markers in each column are pushed by the mercury to record the displacement limits.

Measuring Wind Speed – Anemometer

On bridges, at airports and on many public buildings it is possible to see an instrument with rapidly revolving small cups attached. This is the most common type of anemometer, an instrument used to measure wind speed. The word anemometer derives from the Greek word 'anemos' which means wind. The first anemometer was invented in the 15th century by an Italian, Leone Battista Alberti. Two basic types exist: the velocity anemometer and and the pressure anemometer.

By far the simplest device is the cup anemometer mentioned above. It was invented in 1841 at an Irish Observatory by Dr John Robinson. Three or four small hemispherical cups are attached at each end of two horizontal arms which are in turn mounted on a vertical shaft. As the wind blows it is caught in the cups which turn the arms in a way that is proportional to the wind speed. While simple to use, it is relatively ineffective without electronic data-logging equipment attached. Regular reading and recording from the anemometer can provide a trend but will be likely to miss anything between readings such as a strong gust.

An extremely delicate but highly accurate method is the hot-wire anemometer. It uses a very fine wire, commonly of tungsten, heated to some degree above the ambient temperature. Wind flow past the wire has a cooling effect on the wire. Using this it is possible to calculate the speed of the wind.

Left: An anemometer – an instrument used to measure wind speed – located on top of a tower at the Jodrell Bank radio observatory near Manchester, England. The large telescope in the distance is the giant, 76 metre/250 foot diameter Mark 1A radio telescope. When the wind exceeds a certain speed, the telescope is pointed straight up (as seen here) and secured.

Opposite above: Stevenson Screen containing thermometers and other meteorological measuring and recording devices. The screen is designed to protect the thermometers from the effects of direct radiation and rainfall whilst allowing the free through movement of air. The screens also ensure that conditions at different weather stations are as nearly standardized as possible, allowing a true comparison to be made between temperatures measured at different locations. The thermometers seen here, which are very sensitive to variations, are a max–min set of thermometers (horizontal) and a humidity thermometer (vertical, centre). The devices at either side are recorders.

The Stevenson Screen

No self-respecting amateur weather forecaster would be able to take accurate measurements unless he possessed a simple white box called a Stevenson's Screen. It was designed and invented by the father of the author of *Treasure Island*, Robert Louis Stevenson.

The white-painted, louvre-sided box sits on legs 1.25 metres/4 feet above the ground to avoid interference from the ground. The louvred sides allow a stable airflow through it and over the instruments it contains. It is painted white to reflect incoming radiation from the sun and has a door which opens to the north in the Northern Hemisphere and the opposite in the Southern Hemisphere, in both cases to avoid direct sunshine affecting the readings taken by the instruments inside. Inside are a number of crucial instruments for recording and then forecasting the weather.

Dry bulb, wet bulb and maximum–minimum thermometers are enclosed and are read in that order. The dry bulb thermometer is mercury-filled, recording changes in temperature in degrees and tenths of a degree centigrade. The wet bulb thermometer has a muslin wick covering the bulb. The comparison between the dry and wet bulb temperatures allows the dew point and so the relative humidity to be calculated. The maximum–minimum thermometer contains mercury, which moves up the tube as the temperature increases. A small metal pointer is inside the tube which moves to, and remains at, the maximum or the minimum temperature recorded over any time period before being reset.

The box also contains a barometer which is used to measure air pressure, a critically important part of weather forecasting.

Although many weather stations now use electronic 'remote' sensors, the Stevenson's Screen remains an invaluable part of the weather forecaster's tools to provide accurate details.

Left: Portrait of the Italian physicist Evangelista Torricelli (1608–1647). Torricelli worked on the dynamics of falling bodies with B. Castelli and this led to him being appointed assistant to Galileo, whom he succeeded as mathematician to the court of Tuscany. Torricelli was interested in pure maths and experimental physics, but he is best remembered for his invention of the mercury barometer and his discovery of atmospheric pressure in 1644. He later noticed that small variations in the height of the column of mercury supported by atmospheric pressure were related to changes in the weather.

Measuring Air Pressure – Barometers

A common sight in many homes is the barometer hanging on the wall, usually in a hallway; however, relatively few people know how to use one effectively. Nonetheless, many know that it is used to measure air pressure and that it helps to indicate short-term weather trends. Gases in the Earth's atmosphere have a weight and these exert a pressure on the Earth's surface. It is this pressure that is measured by a barometer. If air pressure is rising this indicates a trend towards more settled weather. When the pressure falls the likely result will be unsettled conditions, perhaps with some rain. Air pressure is most commonly measured in millibars but it can also be measured in pascals or inches of mercury.

In 1608 the Italian Evangelista Torricelli discovered the principles of the barometer by inverting a long glass tube filled with mercury into a dish. Some of the mercury remained in the tube and he realised that the height of the column of mercury varied according to atmospheric pressure. Some people still refer to the term 'mercury rising' when discussing changing air pressure, since mercury barometers are still common today.

During the 19th century the FitzRoy barometer was developed and widely used in ports around the coast of the British Isles. This was the first barometer that could be mass produced at relatively low cost and was used by naval Admiral Robert FitzRoy at the Board of Trade to produce a basic weather forecast for fishermen, the Royal Navy and the merchant marine. Now highly collectible, these are still used today.

Most of the barometers found in homes are the aneroid type. These were first produced by Lucien Vidie during the 19th century. They contain neither mercury nor any other liquid; instead they have a corrugated metal capsule with most of the air removed, which contains an internal spring to prevent it collapsing. As the air pressure rises or falls the capsule sides move in or out, with the movement showing up on a surface dial on the face of the barometer. Whatever method of measuring air pressure, it is usual to record the readings as a trace on a specially designed sheet of graph paper with an instrument called a barograph. A small needle-like pen rises and falls as the air pressure does leaving a continuous trace, a line, on the sheet of paper.

It is worth noting that the best day to set your home barometer is one with calm anticyclonic conditions because it is unlikely that the pressure will change rapidly. It should be set to sea level and as pressure drops by 34 millibars/1 inch per 300 metres/1000 feet of altitude it may need to be adjusted. This is best done with reference to the most local meteorological office.

Sunshine Recorder

A simple glass sphere is used to measure the duration and intensity of sunshine. This is a device known as the Campbell-Stokes sunshine recorder. The sphere focuses and magnifies light from the sun onto a specially treated fire-retardant card and a trace is burnt onto it if the sun is bright enough. The more sunshine there is, the longer the total length of the line.

Rain Gauge

This is used by meteorologists to collect and measure the amount of precipitation that falls over a set period of time. A separate snow gauge measures snowfall. Rain is measured in millimetres or inches, and very occasionally, in centimetres. There are numerous different versions, the most common type being the graduated cylinder rain gauge. It is this type that is usually found in amateur as well as many professional weather stations.

Hygrometer

This instrument measures the moisture content or humidity of the air. The first hygrometer was designed by a Swiss scientist, Horace Benedict de Saussure. The most common type of hygrometer is the dry and wet bulb thermometer. The dry bulb thermometer is mercury filled, recording changes in temperature in degrees and tenths of a degree centigrade. This is the temperature shown on television and radio forecasts. The wet bulb thermometer is an ordinary one with a muslin wick covering the bulb. This is fed by water from a special reservoir attached. The comparison between the dry and wet bulb temperatures allows the dew point and so the relative humidity to be calculated.

Above: The rainfall pollution monitor collects both dry and wet deposition. The dry collector has chemically-impregnated filters to collect nitric acid, sulphur dioxide and particulate pollutants. The wet collector collects rain, and other precipitation, for analysis for pollutants.

Above: A modern sun recorder used to measure sunshine duration. It uses photo-diodes that alter an electrical signal depending on the amount of sunlight present. Electrical heaters ensure dew, frost and ice do not interfere with the readings.

Radar

In making short-term forecasts, radar has become an indispensable part of the weather forecaster's armoury. Modern radar systems are able to produce highly detailed three-dimensional pictures of weather systems such as thunderstorms or fronts and even hurricanes, every ten or fifteen minutes. Radar is the only instrument that can warn forecasters of the imminent formation of thunderstorms or tornadoes. In the USA radar is used to track the eye of hurricanes so that coastal communities have the most accurate information, such as when to evacuate, prior to a land strike. It is now possible to look at live radar images on the internet and via television forecasts.

Radar was first developed during the Second World War to help detect incoming enemy aircraft at the time of the Battle of Britain in 1940 and RADAR stands for 'radio detection and ranging' Once it was realized that it could also detect rain it became an indispensable tool for forecasters.

Radar works by transmitting pulses of radio waves from an antenna. When it meets a 'target' such as a rain shower some of the beam is reflected back.

A Doppler radar measures the change in frequency of the returned echoes which allows for wind speeds to be calculated. Radar devices such as these have a very narrow beam which, when scanning the atmosphere at different angles, can produce a three-dimensional picture of the weather.

The USA has the most comprehensive system of weather radar in the world which is used to monitor the 10,000 violent storms, 5000 floods and as many as 1000 tornadoes it experiences every year. They use aircraft such as the Electra L188L to get as close as possible to the centre of developing storms and so provide a detailed three-dimensional image of the wind and rainfall inside it.

Below: Radar image of a storm. The computer screen shows the output from a Doppler radar, a system used by meteorologists to monitor storms. The different colours represent different intensities of rainfall – a scale is shown on the right. Values of up to 30 DBZ indicate light rainfall and between 30 and 60 DBZ indicate heavy rainfall, whereas anything above 60 DBZ is likely to be hail.

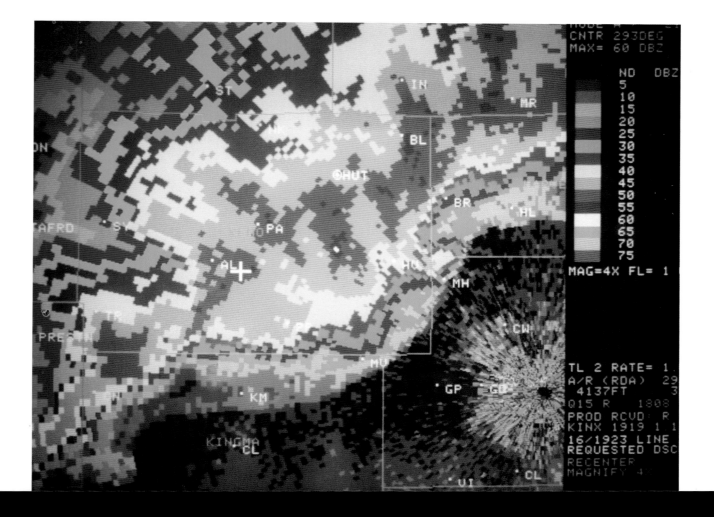

Satellites

Since the launch of the first satellite in the 1950s there has been the exciting possibility of using them to monitor and help predict the weather. The first weather satellite was the 1960 TIROS 1, followed in 1966 by the first equatorial orbiting satellite, essential for monitoring severe weather such as hurricanes. Today satellites such as the METEOSAT geostationary or global orbiting satellites are an essential part of the modern forecaster's tools. They provide a continuous feed of data from all parts of the globe, including those that in the past had been too remote or inhospitable such as deserts and oceans.

There are two main types of satellites in common usage today. The geostationary satellites are positioned at a height of 35,750 kilometres/22,214 miles above the equator and remain at the same spot above the Earth's surface all the time. Meteosat is the geostationary satellite operated by European countries and is positioned at the intersection of 0° latitude and 0° longitude – in other words, where the Greenwich Meridian intersects the equator. From this vantage point it monitors Africa, the Middle East, Europe, much of the Atlantic and the western Indian Ocean.

The second satellite is the Polar Orbiting type, which orbits the Earth via the Poles. The NOAA satellites operated by the USA orbit at a height of 830 kilometres/516 miles, taking about 1 hour and 42 minutes to complete each orbit. The satellite views a different part of the Earth during each orbit as, whilst on its journey, the Earth has rotated through 22°. During 2006, Metop, a European satellite will replace one of NOAAs. With such a low orbit the satellite can record far more detail than that of the geostationary type.

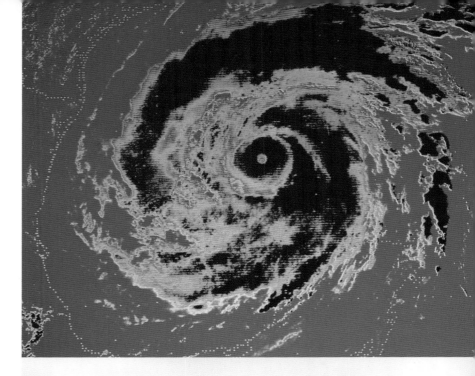

Above: False-colour weather satellite photograph of Hurricane Allen over the Gulf of Mexico.

Below: Technicians working on the Meteosat 6 weather satellite at the Aerospatiale test facility in Toulouse, France. A full-disc image of the Earth is returned every 30 minutes from its geostationary orbit.

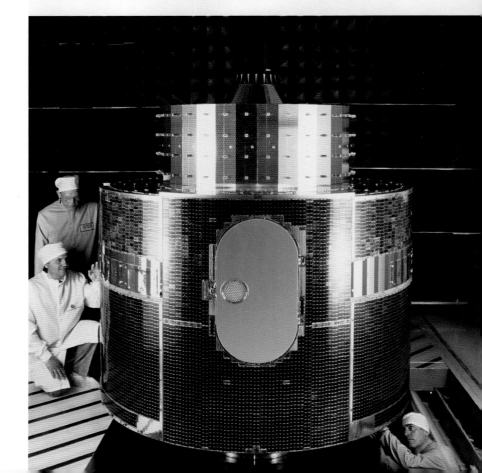

Satellites contain instruments to measure a variety of parameters of use to forecasters and in climate modelling. Some provide the images which everyone will be familiar with from television weather forecasts. These are known as radiometers. Other instruments measure humidity – interferometers – and temperatures – spectrometers. Other sensors are employed, such as a form of radar that can be used to measure cloud cover or rainfall. Satellite images are available through a number of different channels such as visible or infrared. The visible images are rather like conventional black and white photographs and, like photographs, can only be taken when there is daylight. Infrared images however, can be taken even when it is dark.

All satellites are useful in monitoring and tracking particular aspects of the weather, from the intense and dangerous hurricanes or tropical cyclones to more routine high and low pressure systems. They can also be used to measure wind speed and analyse cloud patterns.

Without a doubt, the development of modern satellites has become an increasingly important tool for both forecasting the day-to-day weather and in helping to make climate predictions. The newest generation of satellites being launched during the early years of the 21st century may prove invaluable if predicted changes in global climate prove true.

Opposite main picture: Researcher checking an automatic weather station (AWS) on Alexander Island, Antarctica. The AWS is unmanned and records basic information such as wind speed, daylight hours, temperature, humidity and atmospheric pressure. Data is relayed via satellite to a base station for analysis.

Left: Radar image of a storm cloud (red and yellow) This radar is used to monitor an armoured aircraft as it flies through a thunderstorm collecting research data.

Climate Modelling

In the UK the main centre for climate research is the Hadley Centre, a branch of the Meteorological (Met) Office. Similar research establishments exist in many countries around the world. In each, powerful computers are used to predict or model how weather and climate components may alter when different variables are taken into consideration. Given the complex nature of the dynamics of the atmosphere this may be necessary if predictions are to be of any value. The only alternative to complex computers is a much less complex model and this may be of far less long-term benefit.

At the Met Office the NEC SX-6 super-computers run the highly detailed three-dimensional representational models that are used for the majority of the work.

One type of climate model used is the Atmosphere General Circulation Model (AGCM). This consists of a three-dimensional model representing the atmosphere linked with the land and the cryosphere. Unlike a similar model used for numerical weather prediction – forecasting – these project as far ahead as decades or even centuries and so the level of detail in the variables inputted is low. They are useful for studying the general atmospheric circulation, the variability of climate and how it responds to sea-surface temperature changes.

Other climate models include ocean general circulation models, carbon cycle models, atmospheric chemistry models and regional climate models. These all deal with different aspects which influence climate but perhaps the most complex of them all are the coupled atmosphere-ocean general circulation models (AOGCMs). These are sometimes used to predict the rate of change of future climate.

Given the enormous processing capacity required to deal with a massive amount and range of data, it isn't always possible to deal with it all in one computer. An innovative scheme introduced in 2005 in the UK has allowed home computer users to make their spare processing capacity available to run different programmes. Thousands of people had signed up to take part in this by the summer of 2006. How useful the results will be in enabling future climate components to be mapped and for changes to be predicted remains to be seen.

Left: Computer simulation of a weather map which forms part of the FASTEX (Fronts and Atlantic Storm Tracks Experiment) weather study. The solid lines show isobars which link areas of equal pressure; the closer the isobars appear on the map, the higher the winds will be in these areas. The FASTEX study uses data from aircraft, ships and satellites to discover what factors turn a normal low-pressure weather system (cyclone) into a storm.

Opposite above: Computer graphic showing weather patterns in three dimensions. The three-dimensional weather patterns (upper frame) have been processed from data collected by an airborne double-beam Doppler radar which monitors air movements. In the lower frame is a map of air pressure, with lines (isobars) linking areas of equal pressure.

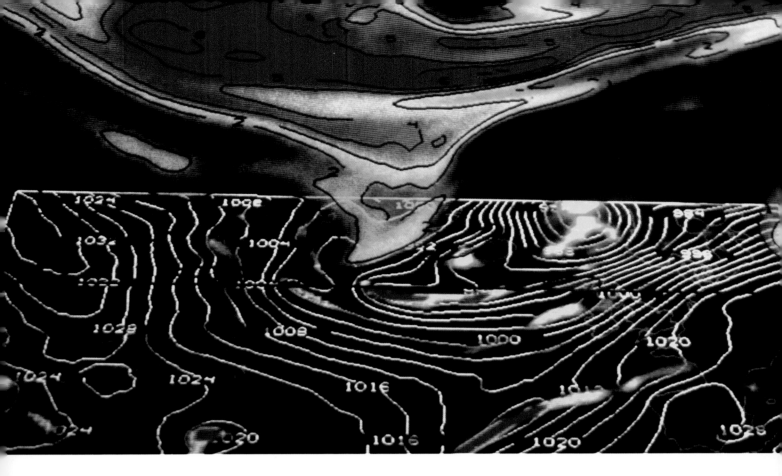

Forecasting and Mapping

Anyone of a certain age with an interest in the weather will surely remember the television and newspapers using maps covered with what at first may have seemed like meteorological hieroglyphics. These maps were widely used at one time and are more properly called synoptic charts. These are still used by weather forecasters today but in tandem with remotely sensed electronic data processed on complex computer programmes.

A synoptic chart shows the weather for a particular area at a specific time. It is the result of the collection of a mass of data from numerous weather stations. These are refined, usually by computer, and plotted using a range of internationally accepted weather symbols.

There are several different types of synoptic charts. The most sophisticated is one which has a huge amount of detail covering, for example, the amount of cloud of various types at different altitudes; dew point temperatures and pressure trends. These are highly specialized and would normally be used only at a national or regional weather forecasting centre Before the use of computer drawing such a map was considered something of an art, which some aerographers (weather men) in the US Navy continue to practice.

At a more recognizable scale a synoptic chart for a specific weather station will display selected meteorological characteristics. These are usually temperature, pressure, cloud cover, present weather – is it raining, for instance – and wind direction and speed. This type of map was commonly seen in newspapers and sometimes on television in the past.

Today, however, the daily weather map is the one people are most likely to come across in the media. This aims to give a clear, if highly simplified and general, impression of the weather. Such synoptic charts are today almost always seen in conjunction with satellite images of the same area.

Despite the array of information at their disposal forecasters can still claim only as much as 70 per cent accuracy in their forecasts shown on these maps. Weather data is a sample taken at different locations rather than a total picture. As such there is always the risk of the anomaly becoming the reality for us on the ground. It may be worth keeping an umbrella handy!

Met Offices

All around the world people depend upon an accurate weather forecast, from the ordinary citizen to more specialist users such as farmers, shipping or aviation companies. These forecasts depend upon accurate and extensive data collection which is both organized and processed by national meteorological offices around the world. Virtually every country has a national meteorological service, from the Australian Bureau of Meteorology to Pakistan's Meteorological Department. Some countries have, in addition to a national service also have additional forecasting institutions such as in the USA where there is the National Weather Service in Maryland as well as organisations such as the Weather Information from the Department of Meteorology of the University of Hawaii. In the United Kingdom the Meteorological Office (Met office), now based in Exeter, has existed since the 19th century, originally developing as a branch of the Board of Trade's forecasting service which began under the guidance of Admiral FitzRoy in the 1860s. It, along with all the others, now relies very heavily on the enormous range of data collected from around the country, from at sea, and via other meteorological offices. This data is entered into computer programs which are used to help predict short and longer term weather. The Met Office has one of the world's largest computers to process the enormous quantities of data collected every day.

The type of forecast given will vary from country to country and will provide not only the basic forecasts for the country and regions of different scales, but can provide specialist forecasts. In the USA the National Weather Service provides detailed forecasts for a number of events such as flood, freeze, gale and storm warning services, a number of different snow forecasts, a high wind watch service, lake and sea surf forecasts and a special winter advisory service. It is possible, via their web site, to find the likelihood of bush fires and hurricanes, along with forecasts for the marine and aviation industries. Most countries' meteorological services are now available via their websites and provide an enormous variety of current and past forecasts as well as a huge number of specialist services.

Weather forecasts often depend on data from other countries and international cooperation is increasingly important. It is not a new phenomenon: in the mid-19th century Admiral FitzRoy of the British Royal Navy began his first rudimentary forecasts in the 1860s. He exchanged data and methodology with the fledgling American Meteorological Service led by Captain Matthew Maury as well as the newly developing meteorological services in France and Germany. International cooperation is now an essential part of predicting not only the weather but future climate. World Weather Watch is one such organisation. The World Meteorological Organisation, with 187 members, is another and is a branch of the United Nations. As the weather knows no boundaries, and as concern about global climate change grows, organizations such as these will surely play an increasingly important role in the future.

Above: Weather forecasters monitor atmospheric conditions around the world to predict future weather conditions.

Right: Historical photograph of Ben Nevis Observatory, the first high-altitude meteorological station in Britain, which opened in 1883. Ben Nevis is the highest mountain in Britain, standing 1344 metres/4410 feet tall. The observatory was set up to measure the weather at altitude, and encompassed measurements of temperature, wind speed, rainfall and atmospheric pressure. The observatory was closed in 1904 as a result of a lack of funds.

Above: View of the main building (background) and instrument domes of the Mauna Loa atmospheric research station in Hawaii, USA. The open dome houses solar radiation instruments which measure the effects of atmospheric dust on sunlight. The observatory monitors all atmospheric constituents that may contribute to climatic change, such as greenhouse gases and aerosols, and those which cause depletion of the ozone layer. All these observations are studied and compared to measurements at other locations to detect global trends in the atmosphere.

Weather Broadcasting

It's almost certain that we all take for granted the weather forecast that appears on television, the radio, in newspapers or on the internet; it has become so much a part of our daily lives. It is perhaps the one aspect of science that has been around for ever. Early man must have looked at the skies and realized that the dark clouds represented forthcoming rain. Even Hippocrates, father of modern medicine, said that 'whoever wishes to pursue the science of medicine must first investigate the seasons of the year and what occurs in them'. The science did not get underway until it was possible to measure the elements during the 16th and 17th centuries.

The modern weather forecast began during the mid-19th century when the British Meteorological Office, a branch of the Board of Trade, was established in 1854. It was supervized by Admiral Robert FitzRoy, former captain of *The Beagle* which carried Charles Darwin and his crew. During Fitzroy's time as a naval captain he recognized a number of weather patterns from measurements he took. Using this data he developed a barometer which was used to predict changing weather. The Met Office started issuing gale warnings to shipping in 1861. Some rudimentary forecasts were published in the London *Times*. Formal weather forecasts were published widely in the press from 1879.

In November 1922 the BBC broadcast the first radio weather forecast with regular broadcasts beginning in March of the following year. On 11 November 1936 the BBC transmitted the world's first ever televised weather forecast.

Since then weather forecast broadcasting has been transformed. World-wide television and radio services produce regular forecasts of varying lengths and degrees of detail.

Past Climates and Trends

Around 4600 million years ago our Earth was formed. At some point up to about 2700 million years ago glaciers and ice sheets began to develop. Since then the Earth has undergone numerous minor and major climate shifts. The last 65 million years is known as the Cenozoic Era; during most of this time the Earth has undergone a period of cooling, albeit rather irregular in pattern. The part of the Cenozoic Era that we are currently in is known as the Quaternary geological period which began about 1.6 million years ago. The first part of this, the Pleistocene period, saw seven major glaciations during which as much as 30 per cent of the Earth was covered by ice at any one time. The current part is the Holocene Epoch which is a time of warm temperatures that began about 10,000 years ago. It is possible that this is an interglacial period, one that exists between glacial periods. The mass of evidence that suggests the Earth's atmosphere is warming due to human action may well be true but we ought to be aware that we have seen fluctuations like this in the past. It is highly possible that, despite the apparent evidence, we still do not yet know what lies ahead for our climate.

The Palaeozoic – The Pre-Dinosaur Period

Around 370 million years ago the first sedimentary rocks such as limestone, sandstone and chalk were laid down under the sea or at the bottom of lakes. At this time the Earth's temperatures were about 10°C/18°F higher than today. The first life forms, algae, were formed about 3500 million years ago. From that time onwards until the Mesozoic Era – the time of the dinosaurs – about 250 to about 65 million years ago, the Earth warmed then cooled repeatedly. This led to periods of intense glaciation which were followed by periods of warmer conditions. We have evidence of this from the sedimentary rocks laid down during these periods. Plant fossils preserved in the rocks indicate that the climate during the pre-dinosaur period was much warmer and more humid than today. Much of the fossil evidence remaining is of ferns; these plants need warm and humid conditions to survive and their remains, laid down first in the Devonian and also in the Carboniferous period (from 345–280 million years ago) provide the evidence of that. Many of the coal deposits laid down in the Carboniferous period and mined during the 19th and 20th centuries contain fossilised seed ferns.

The Mesozoic – The Dinosaur Period

The film *Jurassic Park* and its sequels reflect our continuing interest in the dinosaur period, the Mesozoic Era. The Triassic and Jurassic geological periods saw the evolution of a wide range of dinosaur species made possible by the presence of warm seas. Such sea conditions are usually associated with higher rainfall totals over the land which can encourage lush vegetation growth and in turn encourage the development of huge herbivore dinosaur species. Towards the end of the Triassic period the Earth was subjected to massive upheaval caused by continental drift – the gradual 'drifting' of the Earth's continents across the globe. It would have had major effects on global ocean currents and on continental temperatures. It is speculated that this may have had an effect on growing conditions and the availability of food for the dinosaurs and so was perhaps responsible for their extinction.

The Pleistocene Period – Recent Ice Ages

Periods of glaciation – ice ages – have been regular occurrences in the Earth's history. The evidence suggests that these occur approximately every 200 to 250 million years on average and that they have lasted for millions of years. A period of glaciation consists of major climate differences between the poles and the equatorial regions. During an ice age the ice at the poles extends into lower latitudes creating extensive ice sheets across the land and the sea as well as the development of huge valley glaciers. In this, the Quaternary phase of the Cenozoic Era, the Earth has experienced three significant periods of cooling at about 36, 15 and 3 million years ago.

Opposite: Fossil remains of Archaeopteryx. It was found in the fine-grained limestones of the late Jurassic period (195–135 million years ago) at Solenhofen, Bavaria. Archaeopteryx, seen as the link between reptile and bird, had many features of a small dinosaur with the addition of feathers and combined reptilian and bird-like claws on its wings. It was approximately the size of a crow.

Right above: View of a cliff of sedimentary rock in Zion National Park, Utah, USA, showing very distinct strata. The upper layers of the rocks were deposited about 10 million years ago, whilst the lowest layers are over 250 million years old. The rocks are different colours due to their composition: most of the red and yellow rocks are sandstones with differing amounts of mineral inclusions. The prominent grey layer is from the Chinle formation, a graywacke deposit from the late Triassic period of about 210 million years ago, in which dinosaur footprints have been found.

Above: Fossiled remains of an Eocene Age fish (*Diplomystus dentatus*) some 50 million years old, found in Wyoming, USA.

Right: Glacial erratics at Dusty Basin, King's Canyon National Park, California, USA. An erratic is a boulder which has been transported from its original source by a glacier, and has been left stranded by the melting of the ice. Erratics can be carried a considerable distance and are therefore often of a different type from surrounding rocks.

Below: Ancient Earth; computer artwork of the Earth at the time of the death of the dinosaurs, some 65 million years ago. The extinction of the dinosaurs is thought to have been brought on by a massive asteroid impact on what is now the Yucatan Peninsula in Mexico. This event marks the boundary between the cretaceous and tertiary eras, and is known as the K/T boundary. North is at top. Land is green, water is blue and clouds are white. The continents of North America (top centre), South America (lower centre) and Africa (centre right) are seen. The position and shape of the continents has changed over time due to continental drift, erosion and climate change.

There is a huge array of geological evidence to show that ice once covered huge areas of the Earth that are now ice free. Deep, steep-sided U-shaped valleys, often with mounds of debris called moraine on their floors and at their sides were left behind when the ice melted; these can still be seen today in places such as northern Europe and parts of North and South America. It is possible to identify glacial processes currently operating in place that still have ice today. The resulting landforms can then be seen in other parts of the world and can provide some of the evidence for climate change. The first person to propose that a sudden cooling of the Earth's climate in the past had led to a period of glaciation was a Swiss scientist, Louis Agassiz in 1837. He came across huge granite blocks, called erratics by geomorphologists, during a visit to the Jura Mountains. He felt sure that they could only have been transported there from their point of formation by something as powerful as a glacier.

Although the recent evidence suggests that almost all of the world's current ice sheets and glaciers are retreating (melting) there remains enough uncertainty as to the cause. Whether it be anthropogenic warming of global climate or part of a natural cycle of regular climate change is yet to be fully understood.

Milankovitch Cycles

Serbian astrophysicist, Milutin Milankovitch dedicated his career to developing a mathematical theory of climate based on the seasonal and latitudinal variations in solar radiation received by the Earth. The resulting Milankovitch Theory states that there are cyclical variations between the Earth and the Sun combine to produce variations in the amount of solar radiation reaching the Earth.

The Earth's orbit, already slightly elliptical, is believed to change shape at regular intervals. At some period of time the orbit around the sun is more elliptical and so it receives less solar radiation.

The angle at which the Earth's axis lies with the plane of orbit is currently 23.44°. Milankovitch believed that this angle changes periodically and that it could be between 21.5° and 24.5°. This would also affect the amount of solar radiation received by the Earth and so lead to warming or cooling cycles.

Milankovitch also identified 100,000 year ice age cycles of the Quaternary Glaciation period over the last few million years. He believed that there would be one full period of precession every 26,000 years, in other words there would be a change in the direction of the Earth's axis of rotation during that time. This behaves in a similar way to that of a spinning top as it winds down. It traces a circle on the celestial sphere over a period of time.

Taking all three ideas together the periods of these orbital motions have become known as Milankovitch cycles. They are now used to try to understand the cyclical changes of the Earth's climate.

Below: Global ocean circulation; satellite image of the Earth from space with a map of ocean circulation. Ocean currents flow around the world due to differences in temperature and salinity. A current of warm water (red) from the Pacific Ocean travels westward on the ocean surface. As it flows, it evaporates and becomes saltier. The Gulf Stream carries the warm, salty water up along the East Coast of the United States, then towards Europe. At colder northern latitudes, the water becomes so dense that it sinks to the sea floor and travels south (blue). Disruption of the Gulf Stream from global warming could result in ice-age conditions in Europe.

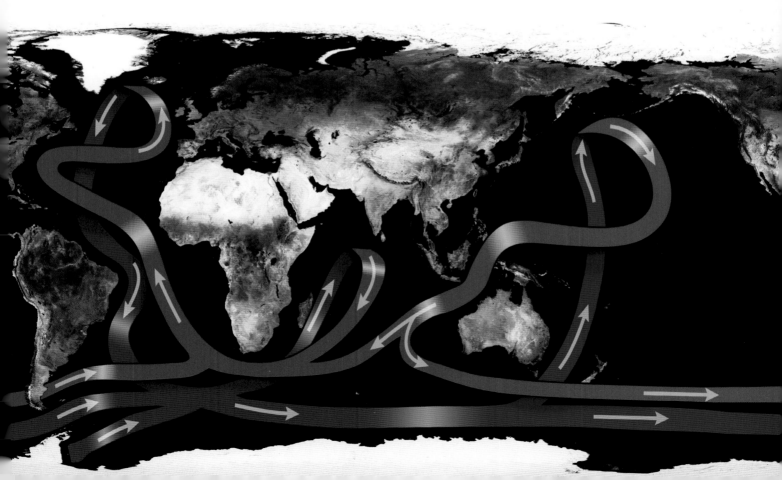

Ice Cores

We have a massive body of organized, historical weather records to act as evidence for changing weather and climate from the last 150 years or so. Evidence of what weather and climate was like in the distant past is rather more elusive, however, but it can be found. The most common source is found in tree rings, in sediment at the bottom of seas and lakes and in ice cores taken from glaciers or ice sheets.

In the Arctic and, particularly, the Antarctic climate scientists have been taking samples of the ice found in the vast ice sheets that cover the surface. These are known as ice cores and are drilled from the surface of the ice deep into the ice at depths of up to 3.22 kilometres/2 miles. At this depth the ice can be up to 100,000 years old consisting of compressed snow which fell at that time. The snow never melts and so the differences between summer and winter accumulation is very evident and this explains why the ice cores are so useful in showing seasonal and long-term climate changes.

Anything else that was in the atmosphere at the time will be deposited along with the snow, such as ash that fell from Krakatoa in 1883 or even earlier deposits from lead smelters in Roman times. Pollen that fell with the snow can show what plant species were common at different times in the past, reflecting variations in climate.

In addition to this, and a very important recent source of information, are the natural variations in dissolved acids and salts along with gases and liquids trapped in the minute air bubbles in the ice. Studying the variations in 'greenhouse' gases such as carbon dioxide is proving a very useful tool in tracking past climates and so in the potential for predicting future changes.

Above: Scientists collecting a 10 metre/33 feet ice core, drilled from the Antarctic near the Rothera Research Station. Chemical analysis of ice cores can reveal changes in climate over several hundreds of years. Air trapped in the firn (the upper layer of frozen snow) is untainted by the atmospheric air, and thus provides direct evidence of what the composition of the atmosphere was at the time when the firn froze. Studying how the atmosphere has changed can lead to better models for predicting how it will change in the future.

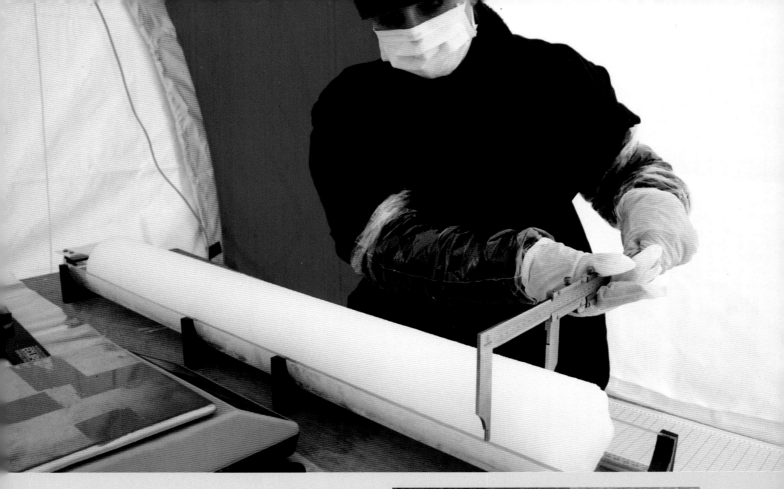

Above: Researcher measuring an ice core to help determine its density, which in turn helps to date the ice. The ice formed over 1000 years ago from the compaction of snow and analysis can provide information about the climatic conditions in which it formed. Cores can reveal changes in climate and atmospheric composition over thousands of years. Knowing how the climate changed in the past allows more accurate predictions of future changes. This core was taken during the EPICA (European Project for Ice Coring in Antarctica) from the ice at Dronning Maud Land, Antarctica.

Right: The Swiss-American naturalist Jean Louis Rodolphe Agassiz (1807-1873) giving a lecture. Agassiz studied fossil fishes with enthusiasm and detail and between 1833-1844 he published all his observations in five volumes. Agassiz first proposed and then verified that glaciers slowly move and also that in the past they occupied vast areas during the existence of an Ice Age. In 1846 Agassiz moved to the United States and in 1861 he became an American citizen. He always opposed the evolutionary theory proposed by Darwin, although his work on fossils helped to establish it.

Other Paleoclimatic Evidence

The study of dendrochronology, or tree rings is one well used source of evidence for past climates. Each year's growth is recorded in a clearly discernable ring which is a record of the climate and growth conditions of that time. It can reflect the precipitation, sun, wind, soil, temperature and snow fall experienced in the area where the tree grew. Variations in the size of the tree rings show when the tree has been affected by, say, drought or lack of sunlight caused by major events such as volcanic eruptions. Such records go back to the end of the last ice age, some 10,000 years ago. The oldest living trees are the bristlecones of California, known to live for up to 5000 years and so providing a major source of these records.

Sediments from sea or lake beds can be extracted by the same process as happens with ice cores. In a similar way the sediments act as a record of the amount and type of material deposited. In a very dry period the layers of sediment are likely to be very thin whereas in a much wetter period the opposite is likely to be the case. The size of the sediment particles can also provide a clue since larger sediments can only be moved if there is sufficient energy in the streams entering the lake or sea to pick them up. In March 2003 an article in Science magazine suggested that lake sediments from Mexico and Venezuela provide the evidence that a series of droughts occurred between 810 and 910 AD. These dates correlate well with the known abandonment of Mayan cities and it has been suggested that the civilization collapsed due to the droughts.

Above: Lone tree on top of a hill called a drumlin, a feature formed by the action of a glacier. This drumlin is near Menzingen, Switzerland, and was formed by the Linth Glacier that advanced over this area in the last Ice Age some 10,000 years ago.

Right: Each year the outer, living part of a tree trunk lays down a new layer of water-conducting xylem cells. The older cells dry out and die, forming the hard mass of wood in the core of the trunk. Each ring corresponds to a single year, so the tree's age can be determined by counting the rings. Climatic fluctuations from year to year result in distinctive patterns of wide and narrow rings, which are repeated in different trees.

Opposite: A glacier is a huge mass of ice formed of compacted snow, moving slowly over a land mass. As a glacier retreats, it leaves behind piles of such rocks, called glacial moraine. A glacier retreats when it melts faster than snow accumulates.

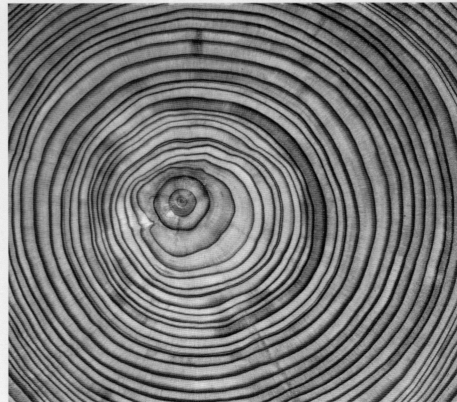

The Little Ice Age

Wide-spread famine, advancing Alpine glaciers and freezing of the river Thames in London – all were consequences of a temporary cooling of the Earth's atmosphere during the 17th and 18th centuries.

The cool temperatures of the period are believed to have shortened the growing season by as much as three or four weeks and may have been responsible for the increase in grain prices by more than five times during this period. This is reflected in the shortage of food produced and the reported famines across Europe during this time. In 1700 a famine was reported in the higher altitude regions of Scotland, the kind of marginal area most vulnerable to climatic fluctuations.

This period of exceptionally cold winters caused the Baltic Sea and major rivers like the Thames to regularly freeze over. 'Frost fairs' were held on the Thames during this time. In 1564 the earliest record of people using the frozen river for entertainment, such as archery competitions, emerges but such activities really took off during the 17th century. In 1685 the Thames was frozen for two months and enterprising businessmen set up stalls selling food and an ox was roasted on the ice. Riverboat men, the taxi drivers of the day, struggled to make a living and resorted to breaking holes in the ice and charging people to cross from the river bank to the solid ice of the frozen river. Even King Charles II and his family visited the fair in this year, reflecting the popularity of the event. More frost fairs followed in 1715–6, 1739–40 and 1813–14, the last 'proper' frost fair. The peak period of the Little Ice Age was 1550–1750.

Gilbert White, a renowned naturalist, wrote in his diaries in January 1776 that the temperatures during that period were unusually low. '...and on the 31st of January, just before sunrise, with rime on the trees and the tube of the glass, the quicksilver sank exactly to zero, being 32 degrees below the freezing point.' He went on to say that it was a 'most unusual degree of cold this, for the south of England'.

It is early documentary evidence such as this that has helped to form a picture of the climate of the time and which supplements other more recent scientific evidence.

Main: Brueghel's painting 'The Census in Bethlehem' from 1566 shows children playing on a frozen river, reflecting the harsh winter conditions of the time.

Inset: A 'frost fair' held on the River Thames in 1683.

Volcanic Effects on the Weather

In an already cold period, the year 1816 is sometimes known as the 'year without a summer'. This was the final winter in which the Thames was frozen over enough to allow a 'frost fair' to take place but this was an exceptional summer. The poet Byron wrote 'Darkness' in 1816, perhaps using the weather as a metaphor for some emotional difficulties. The cause of the poor weather is now believed to be the eruption in 1815 of the Indonesian volcano, Tamboro. During the eruption huge quantities of dust, ash and gas were ejected into the atmosphere. These elements, such as sulphur dioxide droplets, combined to reflect incoming solar radiation back into the atmosphere

and so reduce the temperature at the Earth's surface by as much as 3°C/5°F. It was a time of major weather-related disruption recorded in western Europe and northeast USA. In New England and Canada each of the summer months of 1816 witnessed frosts. In addition the aerosols given off were responsible for the brilliant sunsets witnessed around the world for several years afterwards.

In Iceland, the Laki volcanic eruption of 1783 was one of the largest effusive (lava) eruptions of historical times. It was this eruption that prompted American diplomat and scientist Benjamin Franklin to hypothesize that there was a link between volcanic emissions and a reduction in temperatures.

More recently when Mount Pinatubo in the Philippines erupted in 1991 it ejected enormous quantities of sulphur dioxide – an estimated 22 million

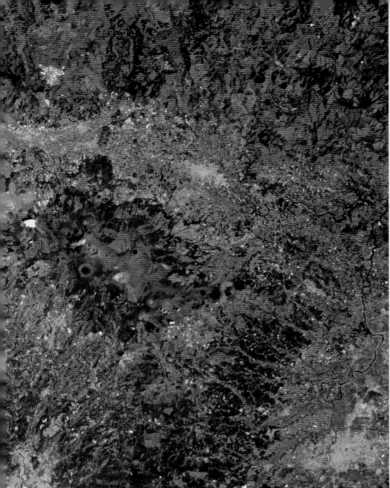

Opposite: The Big island of Hawaii is entirely formed by volcanic activity. Volcanic eruptions can send vast amounts of ash and sulphur dioxide into the upper atmosphere. The sulphur dioxide reacts with stratospheric water vapour to produce a dense haze that can stay in the stratosphere for years. This haze absorbs some incoming solar radiation and reflects more back out to space. This raises the temperature of the stratosphere and cools the lower levels of the troposphere.

Left: Infrared satellite image of the Yaku volcanoes (at centre) on the island of Kyushu, Japan. In the infrared range green vegetation is red, buildings and bare rock are light blue, water is black and clouds are white. The craters of the volcanoes are largely filled with water.

Below: The dramatic landscape of Hawaii has been shaped by volcanic activity.

metric tonnes/24.3 US tons – into the stratosphere where it combined with water vapour to form an aerosol of sulphuric acid. This blocked out the sun's rays sufficiently to cause global temperatures to cool by as much as 0.5°C/0.9°F during the following year.

As recently as the 1980s it was discovered that the Yellowstone area of the USA is a so-called 'super-volcano', a volcano of such an enormous size that when it erupts its effects are likely to be catastrophic, and not just in the USA itself. It has been estimated that a future eruption, given the amount of dust and gases emitted, would lower global temperatures by 5°C/9°F, cause widespread famine and possibly precipitate another Ice Age.

Sulphur dioxide gas, when emitted from a volcano, reacts with atmospheric moisture and can form volcanic smog, 'vog'. This has been well documented in Hawaii where the volcano Kilauea has been erupting for several decades. There it has aggravated respiratory problems, corrodes metal and damages crops. The drinking water of many homes on the island has been contaminated by lead leached from roofs and pipes by the acid rain that falls.

HARNESSING AND MODIFYING THE WEATHER

Renewables

Harnessing the weather to create energy is hardly new. Humans have been making use of the sun, water and wind for thousands of years to create heat, for transport, farming and industry. The Greek island of Karpathos has used windmills for hundreds of years and still does today. The sight of windmills in many parts of the Netherlands is also an iconic image of the country, even if their use is not very widespread today.

As supplies of endangered fossil fuels become more precarious and the concern about the pollution they generate has grown, we are increasingly returning to the idea of using weather-related 'renewables'. The technology of solar, wind and hydro-electricity production has advanced and what were once seen as marginal energy sources are now becoming increasingly important. Of course, all these sources of energy are weather dependent.

To generate enough solar power for heating or electricity production it is essential to have clear skies and bright sunshine. For this reason its large-scale production is currently limited to locations in reliably sunny areas such as California or Southern France.

Wind power needs reliable winds of a reasonable strength to be efficient, which again limits its effective locations to the windier parts of the world such as much of northwest Europe. Wind farms are considered by some to be an eyesore and sites have to be carefully chosen to reduce the perceived impact on the environment.

In a similar way, hydro-electricity can only be generated where there is a suitable site with sufficient reliable rainfall or snow to maintain the water level needed. Developing hydro-electricity plants means that large areas of land will be flooded for ever, altering the environment and often displacing thousands of people in the process.

None of these create pollution in the conventional sense but are often seen as visually intrusive. Wind turbines can be seen from considerable distances and, somewhat inevitably, tend to be located in environmentally sensitive areas which creates conflicts between environmentalists and those proposing the scheme. No-one yet knows for sure what the effects of solar panels are on desert or semi-desert environments but their visual impact is undoubted.

Despite these obvious drawbacks and the attendant controversy surrounding them, there is no doubt that the future of energy production will increasingly depend on the weather.

Cloud Seeding

In 1946 in Massachusetts, USA, an experiment took place to discover whether it was possible to increase the amount of precipitation that falls from clouds. An aeroplane 'seeded' a cloud with crushed dry ice (frozen carbon dioxide) and snow fell from it. A little later it was discovered that silver iodide had the same potential effect and it is now used as an alternative 'seed'.

In areas of the world where rainfall totals are low and where populations are growing, meaning agricultural output has to increase, any means by which rain more can be produced is welcomed. The conventional way of doing this is via irrigation schemes using large-scale dams and reservoirs but in some places the more unusual step of 'cloud seeding' weather modification has been used.

For this to happen clouds containing super-cooled water, liquid water that is below $0°C/32°F$, have to exist. When a substance such as silver iodide or dry ice – which has a crystalline structure similar to that of ice – is introduced it will induce freezing in the cloud. When ice particles form in super-cooled clouds they grow at the expense of liquid droplets and become, the theory suggests, heavy enough to fall as rain from clouds that would otherwise produce none.

As a rule, the cloud seeding elements are dispersed from aeroplanes or via ground-based dispersion devices. Since the first experiments in the 1940s cloud seeding has been used in numerous places. The largest system is in China where they fire rockets into the atmosphere to increase the amount of rain that falls over arid areas of the country, as well as Beijing, the capital. In the USA cloud seeding is used to increase the amount of rain over drought-prone areas, to decrease the size of hailstones that form in thunderstorms and to reduce fog around airports. It is also used occasionally to increase the amount of snow over ski resorts. It has even been used in an experiment to see if the structure of hurricanes would alter. This was abandoned because of the concern from neighbouring areas that they might become more prone to hurricane effects than they were before seeding took place.

Above top: The Russian Myasishchev M-17 was designed primarily for map-making photography and for hail control. In the latter role, the aircraft 'seeds' thunderclouds with small crystals. This causes the cloud to release its rain or hail before it reaches crop areas where the hail could cause major damage. The Russian Committee for Environmental Control also claim that the M-17 would be able to carry ozone generators to 'patch' holes in the ozone layer.

Above: Thunder clouds carrying rain and hail showing anvil formation. A hammerhead or thunderhead is a clear sign of an impending thunderstorm.

Opposite: Traditional windmills have been used for centuries to harness the power of the elements.

Above: Dish-shaped solar power reflectors at a solar power station at White Cliffs, Australia. Each dish focuses radiation from the sun onto a thermoelectric generator positioned in the focal point. A computer steers the dishes to ensure they face the sun throughout the day.

Opposite: Rooftop solar heat collectors (large black panels) that are using the warmth in sunlight to heat water supplies for each house. This saves energy and reduces heating costs. It also avoids the use of power-generation methods that produce pollution, such as burning fossil fuels. Photographed in Stettlen, near Berne, in Switzerland.

Left: Solar-powered emergency telephone being used by a motorist. The solar panel (blue, upper left) converts the energy of sunlight into electricity to power the telephone. The telephone is used to call for help, and the signal is transmitted by the antenna (above solar panel).

240

Solar Power

Humans have used solar power in a very basic form for drying clothes and crops for thousands of years. It is only recently, however, that modern technology has allowed for solar power to be generated almost anywhere, even in high latitudes. Large-scale power stations remain in lower latitudes, however, where the sun's rays are more intense and cloud cover is lower. This means that the amount of insolation, or sunlight received, is high enough to generate either heat, for hot water, or to generate electricity.

There are three ways in which solar energy is generated. The first of these is by use of solar cells – photovoltaic cells – which can convert light into heat to generate electricity. These are used in very sunny areas of the world such as California and southern France. The technology remains relatively expensive and so it not yet used widely in a domestic environment.

Solar water heating via solar panels are perhaps the most commonly used. These use solar energy to heat water for washing, or even for central heating if the sunshine is reliable enough. But more likely they heat the central heating water enough to require little extra conventional heating to get to an appropriate temperature.

A third way is via solar furnaces, a huge array of mirrors which concentrate the Sun's energy into a small space and so produce very high temperatures. The one in Odellio, France has been able to generate temperatures as high as 33,000°C/59,430°F. In California the Solar One power station operates in a similar way but uses the heat to create steam and generate up to 10kw of electrical power in that way.

The advantages are obvious but then so are the disadvantages – at night there is no sun and so no energy. When it is cloudy or in higher latitude areas the energy generated is low and insufficient for anything other than small-scale, low-power uses. This does not mean, however, that in the future most homes will not be able to make use of this technology to supplement or even replace the more conventional energy sources.

Above: Darrieus vertical axis wind turbines (VAWT). Each has four curved blades that are attached at either end to a vertical rod around which they rotate. Heavy equipment like gearboxes and generating machinery can be put near to the ground, reducing the structural support the turbines require. They also work equally well regardless of the wind direction. However, they are not self starting and are vulnerable to damage in high winds. The turbines, owned by Sustainable Energy Technologies, form an experimental 1.5 MW wind power plant in Pincher Creek, Alberta, Canada. This design was proposed by the Frenchman Georges Darrieus in 1931.

Wind Power

The sight of windmills in the countryside of Britain, the Netherlands, Spain or Greece is viewed today as somewhat quaint, traditional and rustic but they show how valuable a source of energy the wind has been in the past and hint at how important it could be in the future.

Early windmills used a system of sails to turn a rotor attached to a series of cogs and wheels. These were in turn attached to millstones which rotated to grind wheat and corn into flour. Sometimes these were used to power the bellows of iron forges. This traditional system is still used in many communities today.

Modern wind farms are a very different prospect. Turbines vary in size from 25–60 metres/ 82–197 feet high with rotor blades spanning up to 65 metres/ 213 feet so their visual impact is significant. Essentially, a wind turbine is an electrical generator mounted at the top of a tower. The wind causes the propeller blades to turn and these rotate a central shaft connected to a generator. This makes them unpopular in some areas: environmentalists complain about the effect on migrating birds and those living near wind farms complain of the noise from the rotating blades. They are, of course, wind-dependent. If there is no wind there is no electricity. Similarly, if the wind is too strong the generators may be switched off. A large number of turbines are needed to generate enough wind to make the farm viable and ways of storing surplus energy are needed.

In 2005 worldwide energy production from wind power was only 1 per cent; however, in Denmark it is 23 per cent and in Spain 8 per cent. Global wind power generation is on the increase: between 1999 and 2005 it more than quadrupled. The proliferation of wind farms around the windy, west-facing coasts of Europe is testimony to the fact that there is huge potential for this source of energy and that its use is being taken seriously.

Due to the controversial nature of their sites on land, numerous off-shore farms have been constructed or are in the planning stage. Small-scale domestic generators are also becoming increasingly economic, efficient and more widely used.

Above: Two types of experimental vertical axis wind turbines, Burry Port, South Wales. Unlike horizontal axis wind turbines this design does not require re-alignment in response to shifting wind direction and there is no need for the generator to be positioned at the top of the tower.

Below: The large sails of the traditional windmill harness the power of the wind by driving the rotation of the machinery inside through a series of cogs. One of the skills of the miller was the adjustment of the blades so that the mill ran at a fairly constant rate.

Left: View of wind turbines on a harbour wall silhouetted against the glow of a sunrise. Each turbine has three blades which rotate around a horizontal axis, turning a shaft which drives an electric generator. The blades have to be turned to face directly into the wind to work efficiently. Wind turbines are an environmentally-friendly source of energy, as they release no toxic waste products and do not rely on a limited resource for fuel.

Below: Wind farms can be situated on land or off-shore. Off-shore sites rely on fairly stable sea conditions, balanced by a reasonable amount of wind. It also allows turbines to be sited away from inhabited areas, reducing their noise nuisance.

Hydro-electricity

At present this is the most commonly used method of generating power from water. Of course, water power has been used for many centuries to generate energy for such things as grinding corn in mills and to power machinery. The first house to be powered by hydro-electricity was Cragside House in Northumbria, UK in 1878. By 1882 the Fox River in the USA generated enough energy to power a house and two papermills. Today, however, it is done on a very large scale, when a dam is constructed which holds back a massive volume of water in a reservoir.

Hydro-electricity is pollution-free and safe, once in place. In the process of construction, however, there can be immense disruption to people's lives and to the environment – villages, even towns can be permanently flooded and wildlife habitats destroyed.

At present a relatively small amount of electricity is generated by hydro-electric power. In the UK it is only 2 per cent but in countries where precipitation totals are higher and suitable sites exist, far more energy is generated. Countries such as New Zealand, Norway and Iceland produce large amounts of energy via HEP.

The Three Gorges Dam in China is, however, perhaps the most famous and, arguably, the most controversial hydro-electric power scheme in the world at present. Located on the Yangtse River near Yichang, construction began in 1993 and was largely completed during 2003. It is predicted to produce about 18 gigawatts to help meet the Chinese population's rapidly growing energy needs when it comes fully on line in 2009. At the outset it was hoped the scheme would provide China with 10 per cent of its needs but such has been the extraordinary growth of the country's economy it is now believed it will be only 3 per cent. The scheme will also improve navigation and reduce the flood risk downstream in the Yangtse valley but has created numerous additional problems for the local population. Almost 2 million people have been displaced due to the flooding of the area to create the vast lake required to store the fuel – water. 40,000 hectares/98,850 acres of the best farmland has gone for ever and the habitat of rare species such as the Chinese River Dolphin has also disappeared for good putting this and other species at risk of extinction.

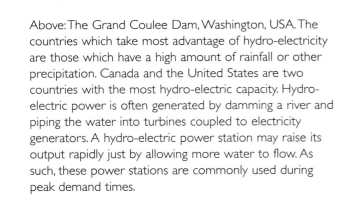

Above: The Grand Coulee Dam, Washington, USA. The countries which take most advantage of hydro-electricity are those which have a high amount of rainfall or other precipitation. Canada and the United States are two countries with the most hydro-electric capacity. Hydro-electric power is often generated by damming a river and piping the water into turbines coupled to electricity generators. A hydro-electric power station may raise its output rapidly just by allowing more water to flow. As such, these power stations are commonly used during peak demand times.

Opposite: Hydro-electric watermill, which has incorporated a turbine for electricity generation since 1888. The technology behind the water mill is somewhat older than the windmill and both the Greeks and Romans are known to have used it.

Right: Water-powered funicular railway at the Centre for Alternative Technology, near Machynlleth, Wales, UK.

FUTURE CLIMATES

Climate Change

The Earth's atmosphere is in a cycle of constant but very slow change. Much well-documented evidence exists to prove that climate has altered many times in the past. This includes glaciated valleys and mountains, sand dunes and geological evidence such as fossils and coal deposits, the relics of ancient tropical forests. The idea of future climate change should not, therefore, come as any great surprise.

Climate change has always been thought to be gradual and slow. It is believed that the last Ice Age ended 10,000 years ago and that it took many centuries for the ice to gradually disappear. It was, until recently, thought that there had been climatic fluctuations over short periods lasting 30 to 100 years but around a relative norm similar to that which has existed for the last 4000 years.

However, in recent years doubt has been cast on these assumptions and climate change is now a matter of great concern and serious public, national and international debate. One of the main issues is that human activity may be responsible for not only a change but also an increase in the rate of climate change. In addition it is now recognized that climate change in the past almost certainly occurred much more swiftly than had previously been thought, in other words in as short a time as a century rather than over many centuries.

The greatest concern now is that these changes are leading to a significant increase in global temperatures. Recent research suggests that the Earth's average temperature has risen by $0.5°C/1°F$ since 1960 and by $0.6°C/1°F$ since the start of the Industrial Revolution. Concern about the effects of global warming has inevitably increased in recent years. Governments have signed treaties such as the Kyoto Protocol of 1997 and had discussions more recently at the Bonn conference of 2006 in order to minimize the amount of greenhouse gases being emitted into the atmosphere in an attempt to reduce the risk of future climate change.

Left: Type II polar stratospheric clouds (PSC II), seen from a NASA DC-8 research aircraft over the Arctic. The bright PCS IIs are at an altitude of 15,300 metres/50,000 feet, and are composed of crystals of water ice. The clouds exhibit a 'wave' structure, caused by the effect of mountains on the winds that move the clouds. PSCs are found in the circumpolar vortex winds that circle both poles. Type I PSCs (made of nitric acid crystals) have been linked with the chemical processes that are depleting the ozone layers over the Arctic and Antarctic regions.

Acid Rain

During the 1970s and 1980s the Scandinavian countries of northern Europe reported significant damage to their coniferous forests and lakes. Forests were dying and the lakes, streams and soil were becoming increasingly acidic. It was believed that the death of the trees was also due to the highly acidic nature of the rainfall over the area. The blame was put at the door of countries further south, particularly the UK, which had been burning fossil fuels for many decades. The prevailing south-westerly winds had blown the resulting gases across the Scandinavian countries where they fell as acid rain.

Acid rain is something of a broad term as it can refer to either wet or dry deposition. What they have in common is a greater than normal concentration of nitric and sulphuric acid. This can occur naturally, such as when volcanoes erupt, but it is the anthropogenic sort, that caused by human activity, that is usually discussed. When fossil fuels such as coal are combusted they release gases such as sulphur dioxide and nitrogen dioxide. These gases react in the atmosphere with water vapour, oxygen and other chemicals to form various acidic compounds. The result is mild sulphuric or nitric acid released in rainfall or acid fog or mist as well as acid snow.

The dry form occurs when sulphur dioxide and nitrogen dioxide are released during combustion and are blown across national borders by the prevailing winds. It is estimated that about 50 per cent of all acidity in the atmosphere occurs in the dry form. These chemicals may become incorporated into the soil as well as into streams and lakes. Some will land upon trees and vegetation where it can be washed off by rain, which it in turn becomes acidic.

Above left: Chimneys of a nickle-smelting factory release fumes into the air in Norilsk, Siberia, northern Russia. The sulphurous fumes settle around the smelting plant as a fine dust which poisons wildlife. Sulphur oxides, once dissolved in atmospheric water, are a major cause of acid rain.

Left: A conifer forest on Mount Mitchell in North Carolina, USA, with many trees damaged or killed by acid rain. Loss of leaves and stunted growth occur as the trees undergo a 'die back' effect. Trees, especially conifers, are particularly vulnerable to acid rain.

The Effects of Acid Rain

Acid rain doesn't directly kill trees and other vegetation; rather it weakens them by damaging leaves which means they cannot easily access nutrients. Acid water dissolves some soluble nutrients and minerals in the soil and leaches them out, in other words it washes them out into lower layers of the soil where the vegetation roots cannot reach them.

Acid rain can also have a noticeable effect on buildings and other structures made of alkaline materials such as limestone. Sir Christopher Wren's St Paul's cathedral in London is built using a form of limestone sometimes called Purbeck marble, the stone is of the very highest order with an almost marble-like appearance. However, the increasingly acidic atmosphere over London has scarred the building: it is estimated that the acid rain and dry acid compounds that fall on the cathedral dissolve the stone to the equivalent of a bucketful each day. The lead plugs used in its construction during the 17th century now stand several centimetres above the surface of the building, reflecting the rapid rate of weathering since its construction. Similar effects can

be seen on churches and other public buildings, even gravestones, caused by acidification of the atmosphere.

The effects of acid rain and its variants on human health are sometimes overlooked but they can be serious. It is the polluting compounds, sulphur dioxide and nitrogen dioxides, in the atmosphere rather than the acidic water itself that are the source of the problem. These gases interact in the atmosphere to form fine sulphate and nitrate particles that can be inhaled deep into people's lungs. Numerous scientific studies have suggested a link between elevated particles and increased illness, even deaths, from heart or lung disorders such as asthma and bronchitis.

Nitrogen oxides are also known to interact with volatile organic compounds to form ozone which impacts on human health causing an increase in the risks associated with lung inflammation including asthma and emphysema.

Above: Relief work on the facade of St Bartholomew's church in midtown Manhattan, New York, damaged from years of pollution and acid rain.

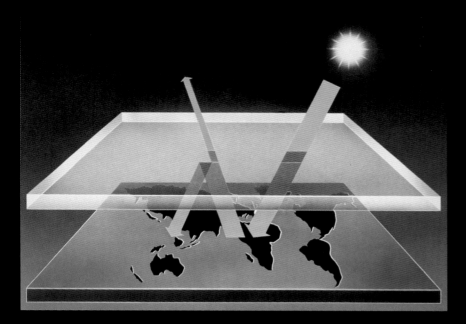

The Natural Greenhouse Effect

The Earth is made habitable by 'greenhouse' gases such as carbon dioxide, ozone and methane. These gases absorb infrared radiation and warm the planet, making it habitable for human and other life. When the balance of greenhouse gases changes a result of human activity, the so-called enhanced greenhouse effect occurs, an idea first explored by Joseph Fourrier in 1824. The Earth receives energy from the sun in the form of radiation, mainly in the visible spectrum. Much of this, about 70 per cent, is absorbed by the land, atmosphere and water on the Earth's surface but eventually much is re-radiated back into space. About 30 per cent of the incoming energy is reflected back into space before it reaches the land. This regular cycle keeps the Earth in balance. If that balance is disturbed the Earth may either cool down or heat up. Minor fluctuation is part of the normal process but it is the threat of major imbalances that is of real concern.

The main 'greenhouse' gases are water vapour, methane, nitrous oxide and carbon dioxide with a significant contribution made by ozone and chlorofluorocarbons. As a result of these processes in the atmosphere the Earth's average surface temperature is kept at about 15°C/59°F. Without them the temperature would be closer to −18°C/0°F, far too low to support life. The atmosphere keeps the Earth warm in a way which is similar, but not identical, to the way a greenhouse keeps plants warm.

Above: Illustration of the 'greenhouse effect'. The build-up of certain gases in Earth's immediate atmosphere - the trophosphere – traps an increased amount of solar radiation and leads to a gradual warming of the whole planet. Here, a polluted trophosphere is represented by the purple band, allowing only a small fraction of re-radiated solar energy (yellow arrows) to return to space.

Below: Diagram showing the chemical composition of air.

Air Composition

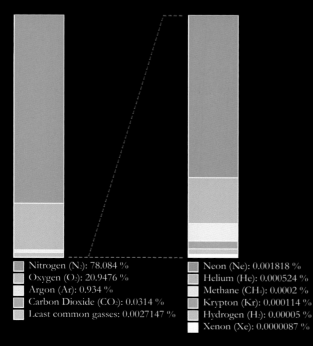

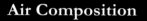

Nitrogen (N₂): 78.084 %
Oxygen (O₂): 20.9476 %
Argon (Ar): 0.934 %
Carbon Dioxide (CO₂): 0.0314 %
Least common gasses: 0.0027147 %

Neon (Ne): 0.001818 %
Helium (He): 0.000524 %
Methane (CH₄): 0.0002 %
Krypton (Kr): 0.000114 %
Hydrogen (H₂): 0.00005 %
Xenon (Xe): 0.0000087 %

The Enhanced Greenhouse Effect

Since the Industrial Revolution there has been a marked increase in the amount of carbon dioxide, nitrous oxide and methane in the atmosphere. Recent data from the Mauna Loa observatory suggest that CO_2 levels have increased from 313 ppm in 1960 to 375 ppm in 2005.

Global climate models suggest that the increase in CO_2 levels may be responsible for a warming of the Earth's atmosphere which it is believed has increased by 0.6°C/1°F in the last 150 years, of which 0.5°C/1°F is since 1960. A considerable body of evidence is being amassed to support the argument that human activity is responsible for this warming but the issue remains controversial.

Above right: Adelie penguins on Torgersen Island, Antarctica. This part of Antarctica has warmed more than anywhere in the world during the 30 years since scientists have been observing the penguins. As a result, the sea ice has retreated and there is more snowfall, making it more difficult for the Adelie penguins to nest and find food.

Opposite: A weather balloon filled with helium is used to carry an ozonesonde, an instrument that detects ozone in the atmosphere, into the atmosphere. Measurements are back to a base station using a radio transmitter and are used to study the depletion of the ozone layer.

The Ozone Layer

Ozone enables the human race to exist yet can make our lives very difficult indeed. Ozone is a gas that is made up of atoms of oxygen. It can exist anywhere in the atmosphere but is mostly concentrated in the stratosphere at around 22 kilometres/14 miles or higher above the Earth. Under certain atmospheric conditions it can also be formed near ground level. Normal oxygen consists of two atoms but ozone has three and has the chemical formula O_3. It is sometimes pale blue in colour and has a quite strong smell. It is also highly poisonous.

Ozone can be detected almost anywhere in the atmosphere but is most common in the stratosphere where it provides a protective barrier against the Sun's potentially harmful ultraviolet rays. Under certain conditions and under certain wavelengths the Sun creates UVB rays which can cause skin cancer. In this case the ozone, sometimes an irritant, is crucial to the well-being of life on the planet.

Right: Computer artwork showing the ozone layer. Oxygen molecules (O_2, upper left) are hit by high-energy photons, causing them to split into two oxygen atoms (photodissociation of oxygen). Individual oxygen atoms (O) can then combine with oxygen molecues (O_2) to create ozone (O_3, centre). The ozone molecules absorb UV radiation (purple beams), which causes the ozone to split into an oxygen molecule and an oxygen atom. They can can then recombine to create more ozone.

Below: Many aerosol sprays are powered by chemicals called chlorofluorocarbons (CFCs).

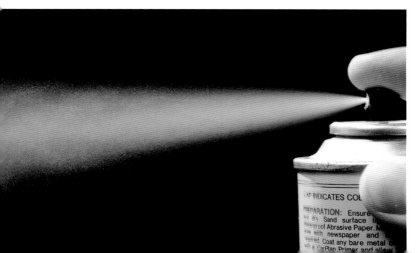

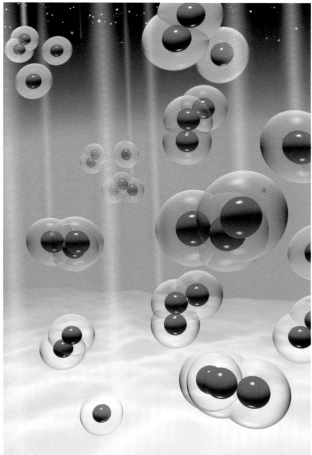

The Hole in the Ozone Layer

During the late 1970s scientists working for the British Antarctic Survey found that the amount of high-level ozone had decreased quite significantly over the continent. The so-called ozone hole is a periodic one that opens in the Antarctic spring and closes several months later. Research carried out in the Arctic discovered a smaller but very similar pattern of ozone depletion occurring there too.

The responsible agent, it was discovered, is a group of chemicals called chlorofluorocarbons (CFCs) used commonly as refrigerants and in aerosol sprays. These chemicals are also used in air conditioning units, cleaning solvents and packaging materials such as polystyrene. The CFCs reach the upper atmosphere where they react with sunlight to produce a gas called chlorine monoxide which is highly reactive and acts as a catalyst for the breakdown of ozone.

In 1985 the Montreal Protocol was signed by 49 countries following a United Nations convention. The idea was to phase out the use of CFCs by the end of the 20th century. In 1989 the EU proposed a total ban on the use of CFCs by the end of the 1990s. Research continues and in 1991 NASA launched a satellite to measure and monitor the variations in ozone at different altitudes with the intention of providing a more complete picture.

Although the amount of CFCs in the atmosphere has been dramatically cut since the 1990s the problem persists; the chemicals have a long life and can remain in the atmosphere for more than 80 years so damage to the ozone layer may last for a long time to come. This means that the Earth is exposed to far more UV radiation that is either necessary or safe. This, it is feared, will increase the incidence of skin cancer and cataracts and reduce the effectiveness of the human immune system. There are also fears that it will interfere with plant photosynthesis and the growth of phytoplankton in the oceans, so affecting marine ecosystems.

Opposite above: Artwork showing the severe depletion or 'hole' in the ozone layer over Antarctica (white landmass). The ozone layer occurs at an altitude of 14–24 kilometres/ 9–15 miles, but is most concentrated at 17–21 kilometres/ 11–13 miles (shown in white). The hole lets through the UV radiation (purple). Altitudes are given by a Boeing 727 (12 kilometres/7 miles), a Concorde (15 kilometres/ 9 miles), and the ER-2 research plane (18 kilometres/ 11 miles) used by NASA to investigate the hole. The hole first appeared in 1980, and has grown more severe each year. It is a seasonal phenomenon that occurs in the Antarctic spring.

Right: Refrigerators and freezers stacked ready for recycling at a refuse disposal site. The machines have their compressors, coolant (refrigerant) and polyurethane foam removed before they are crushed and sold for scrap. The coolants and foam contain chlorofluorocarbon (CFC) chemicals. CFCs are harmful gases that deplete the ozone layer and can trap heat in the atmosphere, which may contribute to the greenhouse effect.

Low-level Ozone

Ozone is present at low levels in the atmosphere, at close to ground level. Although ozone is a toxic gas it is considered relatively safe if it is at quantities lower than 30ppb. It is now quite clear that a number of human activities are responsible for increasing the amount of ozone at ground level to a level where it can be dangerous to human health.

Low-level ozone is one of the pollutant gases that can create significant problems for people with respiratory conditions such as asthma, particularly for the elderly and young children. It is not surprising to find it monitored carefully in order to provide warnings for the vulnerable groups when readings reach critical levels.

It can not only damage human health but affect animals, trees and other plants and is believed to damage rubber, nylon, plastics, dyes and paints. The financial cost of replacing and repairing structures containing these materials can be immense.

Low-level ozone can contribute to acid rain and the enhanced greenhouse effect and is also partly responsible for photochemical smog. Ozone is a by-product of fossil fuel burning. When pollutants such as hydrocarbons and nitrogen oxides react with sunlight they produce ozone. Photochemical smog is a complex pollutant which can be generated when the reaction takes place in strong sunlight which makes the problem acute in urban areas with high sunshine levels, such as in California, and particularly during the summer months. The prevailing winds can blow the pollutants away from urban areas where they are created across to rural areas.

Below: Photochemical air pollution, or smog, over Singapore. It results from hydrocarbons and nitrogen oxides, mainly from vehicle exhausts, reacting together in the presence of sunlight. The pollution limits visibility, damages plants and irritates the eyes.

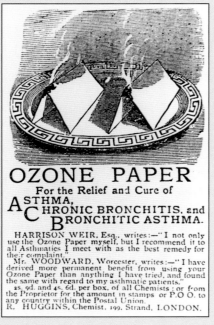

OZONE PAPER

For the Relief and Cure of

ASTHMA,
CHRONIC BRONCHITIS, and
BRONCHITIC ASTHMA.

HARRISON WEIR, Esq., writes:—"I not only use the Ozone Paper myself, but I recommend it to all Asthmatics I meet with as the best remedy for their complaint."

Mr. WOODWARD, Worcester, writes:—"I have derived more permanent benefit from using your Ozone Paper than anything I have tried, and found the same with regard to my asthmatic patients."

2s. 9d. and 4s. 6d. per box, of all Chemists; or from the Proprietor for the amount in stamps or P.O.O. to any country within the Postal Union.

R. HUGGINS, Chemist, 199, Strand, LONDON.

Above: Ozone pollution research. Trees inside a FACE (Free-Air CO_2/O_3 Enrichment) ring for studying the effects of ozone (O_3) and carbon dioxide (CO_2) pollution on plants. O_3 and/or CO_2 are pumped into the air from the top of 2.5-metre/8-feet-tall vertical pipes (white). These trees have been damaged by ozone, which is a pollutant gas formed by chemical reactions between oxygen and pollutants in the lower atmosphere. The levels of ozone are expected to rise with increased industrialization and temperature changes. Carbon dioxide increases plant growth, but ozone reduces it.

Left: Advertisement for ozone paper from the 1880s. Inhalation of the smoke released from burning this paper was claimed to cure and relieve the symptoms of asthma and bronchitis. Ozone, a form of oxygen, is not produced when burning paper, and is poisonous to humans.

Above: Aerial view of the edge of the Greenland ice sheet. The extent of the sheet is determined by the seasons, but a large amount of the most northern areas is always covered.

Global Warming: Evidence

It would be foolish to assume that one or two hot summers or exceptionally cold winters herald the onset of significant climate change but a significant body of evidence now exists to suggest that these freak events are part of a trend. In 2001 the Intergovernmental Panel on Climate Change (IPCC), established in 1988, predicted that global temperatures would rise between 1.4°C/2.5°F and 5.8°C/10°F by the end of the 21st century. They announced that 'most of the warming observed over the last 50 years is likely to be attributable to human activities'. This was the first occasion on which they had drawn the conclusion that climate change and human activity were linked.

Global temperatures have increased by 0.6°C/1°F in the last 150 years with most of that rise being since the 1960s. The 1990s was the warmest decade on record with 1998 being the warmest year since reliable records began in 1861. In fact four out of the five warmest years ever recorded were in the 1990s. In the UK 1995 saw the highest number of hot days – 26 days above 20°C/68°F – in 225 years of daily recording. The total number of cold days (average temperature under 0°C/32°F) each year was between 15 and 20 until the end of the 19th century, but is believed to have fallen to about ten by the end of the 20th century.

Sea levels have already risen globally by 10–20 centimetres/4–8 inches over the last century: sea ice in the Arctic is thinning and rainfall is increasing in some parts of the world. The Greenland ice sheet is thinning, at a rate of more than 1 metre/ 3 feet a year in places. Recent evidence suggests that as much as 51 cubic kilometres/12 cubic miles of ice each year is being lost from the whole of this area. Snow cover has declined by about 10 per cent in the

Northern Hemisphere and mountain glaciers in non-polar regions have retreated significantly during the latter part of the 20th century. Mount Kilimanjaro in Tanzania has lost about 82 per cent of the mass of its ice cap since it was first measured in 1912 and about one third from the top in the last dozen years. Peru's Quelccaya ice sheet in the Southern Andes has lost 20 per cent of its mass since 1963 with one glacier, Qori Kalis, having decreased 32 times faster in the last three years than it did in the period between 1963 and 1978.

In the mid and high latitude parts of the northern hemisphere there has been a 2–4 per cent increase in the frequency of heavy rainfall; totals have increased in those parts of the world in any case, by as much as 0.5–1 per cent each decade. In addition the frequency and intensity of drought in Asia and Africa has increased in the last few decades.

How much of this really proves that major global climate change is taking place remains to be seen.

Questioning Global Warming

Despite the apparently overwhelming body of evidence which supports the theory that the activities of man are having an impact on climate change, there remain a significant number of dissenters in the scientific community.

Their arguments centre on flawed data collection – too many data collection sites are close to urban areas and so may be affected by the heat given off by them. Other arguments relate to inconsistencies between data collected at ground level and in the upper layers of the atmosphere. If global warming is taking place temperatures should increase both at low and at high levels in the atmosphere. Computer modelling may also be flawed as the machines are unable to fully replicate the enormous complexity of the Earth's atmosphere. A final point is that conventional researchers may have underestimated the influence of the sun. Increased solar activity, they believe, will lead to fewer clouds and so more energy reaches the surface thereby warming the atmosphere.

Above: This is how healthy coral should look. The red tentacles of the feeding coral organisms sift food particles from the water. Most corals feed on minute animals or tiny plants and algae.

Below: This coral has been bleached because of the loss of symbiotic algae (zooxanthellae). The exact mechanism or trigger for the bleaching is unknown, but adverse changes in the coral's environment are factors. Increases in sea temperatures caused by global warming and sea pollution have both been cited as probable causes for the death of corals.

Above: It is currently believed that the effect of global dimming may be due to the increased presence of particles in the atmosphere as the particles absorb solar energy and reflect sunlight back into space. One of the causes of this may be the burning of fossil fuels – already deemed to be responsible for pollution, acid rain and global warming.

Left: Smog, a thick, dark, sulphurous fog containing dust and soot particles, lies thickly over a busy street in the city of Buenos Aires, Argentina.

Global Dimming

In 2001 a report was published by Gerry Stanhill, a British scientist, in which he showed that the amount of sunlight reaching the Earth had fallen. Stanhill was working in Israel and had compared sunshine records there from the 1950s with those of the present day and discovered that there had been a 22 per cent drop in the amount of sunlight received. He went on to study the sunlight records from around the world and discovered falls of 10 per cent in the USA, almost 30 per cent in parts of the former Soviet Union and 16 per cent in parts of the British Isles. The amount of sunshine reduction varied from place to place but the global average per decade between the 1950s and 1990s was between one and two per cent. He described the phenomenon as global dimming. Although his results were greeted with some scepticism by other scientists, his conclusions were recently confirmed by scientists in Australia who had used a completely different method.

Global dimming seems to be caused by the burning of fossil fuels which produces carbon dioxide but also particles of soot, ash and sulphur compounds; in other words, atmospheric pollution appears to be the cause. Particles in the atmosphere reflect some sunlight back into space and also change the optical properties of clouds. The particles act as hygroscopic nuclei, the particles around which water molecules collect, and so clouds in polluted areas will contain more water droplets than those in non-polluted areas. According to recent research this makes clouds reflect more of the sun's rays back into space.

There are now suggestions that dimming was responsible for the droughts of sub-Saharan Africa of the 1970s and 1980s. By reducing the amount of sunshine that reaches the oceans, dimming may be disrupting the world's rainfall patterns. The drought in Africa claimed thousands of lives; any future disruption of rainfall could be equally catastrophic.

Paradoxically, some scientists believe that, if this is really happening, it may mean that global warming will be a greater threat to humans than had previously been thought. Some scientists are of the opinion that the Earth has not warmed by as much as expected given the amount of carbon dioxide added to the atmosphere, believing that the warming from greenhouse gases is being offset by the cooling effect of dimming, in other words, one is cancelling out the other. It may be that the world's climate is actually more sensitive to the greenhouse effect than originally thought. Improvements in pollution control are now reducing the amount of particles in the atmosphere at a faster rate than reductions in greenhouse gases. This may mean that the modifying effect of dimming will be reduced in future and that global temperatures will begin to increase faster than previously thought.

Right: Women selling maize, the local staple food, in a market in Galufu, Malawi. It has been suggested that global dimming may be linked to severe drought conditions. In Malawi, and other areas of southern Africa, recent harvest have been poor and future yields are uncertain because of lack of rains. These drought conditions have caused an increase in poverty during the last few years and many areas of southern Africa have been hit by a severe hunger crisis. Malawi is one of the worst affected areas and Galufu village is a typical small village that has become victim of this poverty spiral.

Desertification

The world's great deserts, such as the Sahara, have developed naturally over many hundreds and thousands of years. They have advanced or retreated entirely independently of human activity during this time. There is now some evidence to suggest that the rate of spread of desert areas has increased and that this may be in part due to human activity and global climate change.

Below: Skeletal remains in Sossusvlei, Namibia. Sand dunes are formed from windblown sand. The sand accumulates into ridges or furrows that originally lie parallel to the direction of the prevailing winds. Ripples form as sand is transported and deposited by wind. There are three main types of desert: sandy deserts or ergs, rocky deserts or hammada, and stony deserts or reg. Deserts are made in different ways, but they are always formed because there is not enough water. Only about 25 per cent of the world's deserts are sandy.

Where a desert begins and ends is often difficult to determine and the transitional zone is sometimes gradual. Human activity in these transitional zones may be critical to the formation or extension of deserts. Where human population has increased there may be intolerable pressures on the fragile ecosystems of arid and semi-arid desert-transitional zones. Overgrazing of animals, the cutting down of trees for firewood and poor agricultural techniques may be responsible for degradation of the soil and its ability to support vegetation. Topsoil may turn to dust and be blown away. One prediction for the latter half of the 21st century is that the East African countries of Kenya, Mozambique and Tanzania will experience slightly warmer weather but will be challenged by persistent drought. The inevitable consequences will be a real threat to the food supply from the grain-producing parts of the region. Warmer temperatures will also increase the rate of evaporation in areas already vulnerable, exacerbating the problems people there will have to face in the future. Northern and western Europe are predicted to experience increased soil erosion as rainfall reduces, and the soil dries and is blown away by increasingly strong winds. This currently productive area will be likely to suffer food shortages as a result. The UK is currently only 60 per cent self-sufficient in food production but can afford to import; however, any further reductions will lead to significantly increased problems for the area in future. The unusual droughts there in 1976 and 2006 have hinted at the problems that may become more common as global warming develops.

Above: Rock formation at Golden Gate National Park, South Africa. Windswept sand grains bounce against the rocks, slowly wearing them away. The grains are too heavy for the wind to lift high, so they cut into the rocks near the ground. Some are carved by the wind to look like giant mushrooms.

El Niño

In September 2006 a report stated that sea surface temperatures in the eastern Pacific had risen by 1.5°C/2.5°F in less than two weeks. This, it appeared, was the beginning of an El Niño event. Droughts, floods and increased hurricane activity are all associated with El Niño and with its counterpart in the equatorial Pacific, La Niña.

An El Niño event is a disruption of the ocean-atmosphere system in the tropical Pacific. It brings warm water to the eastern Pacific and is associated with unusual weather across the globe. The phenomenon was initially recognized by fishermen off the Pacific coast of South America who identified unusually warm water at the start of the year. El Niño is a Spanish term meaning little boy or Christ Child, since the weather phenomenon tends to develop around the Christmas period. It is thought to recur every 11 years or so but can develop more frequently: in 1986–87; 1991–92 and again in 1997–98.

Normally the trade winds blow towards the west across the tropical parts of the Pacific. Warm surface water piles up in the western Pacific driven by the trade winds which means that the sea may be 50 centimetres/20 inches higher near Indonesia compared with its level off the Ecuador coastline. Sea surface temperatures are about 8°C/14°F higher in the west than in the cool-water areas of the east.

This area is cooler because cold water from the deep rises to the surface bringing with it many nutrients, which is responsible for the marine diversity of the region. In this easterly area there is relatively little rainfall but there is considerably more in the west as air warmed by the sea rises and is blown to the west across the land.

The situation is different in an El Niño year, however; as the trade winds relax in the central and western Pacific. Sea, surface temperatures rise in the eastern Pacific leading to the increased rainfall normally associated with lands further west. This can bring floods to western parts of South America and conversely less rainfall or even drought over the west in countries like Australia and Indonesia.

The changes are not confined to the Pacific as El Niño is recognized as having an impact on global circulation and an impact on weather around the world. The El Niño year of 2002 was blamed for the severe flooding across Europe and the droughts across the USA, Australia and parts of south-east Asia.

Below: Coloured satellite image of the atmospheric humidity over the Pacific Ocean due to El Niño. The colours run from red (high humidity) through yellow and green to blue (low humidity). Humid air is formed over warm areas of sea because of the large amounts of water which evaporate to form clouds. Here, the water vapour is at an altitude of about 10 kilometres/6 miles.

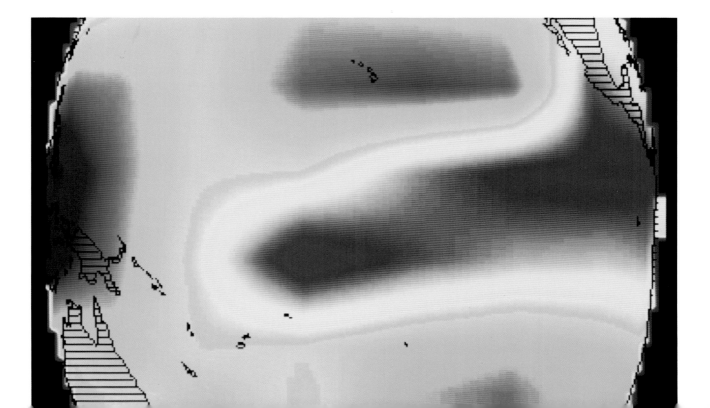

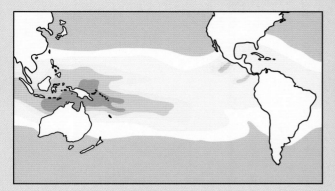

La Niña January - March 1989

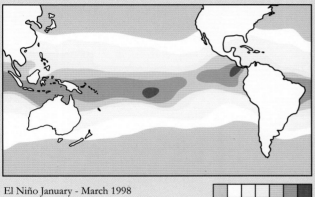

El Niño January - March 1998

20°C   25°C   30°C

Above: Normally trade winds blow west and the sea may be 8°C/14°F warmer in the west than in the east (top), where deep, cold water rises to the surface. In an El Niño year the trade winds relax and sea surface temperatures rise in the eastern Pacific (bottom).

Above right: The name El Niño comes from the Spanish for 'boy child' because in a very mild form this cycle occurs every Christmas off the coast of South America. La Niña is the return to upwelling conditions along the coast of Peru and Chile and heralds the return of plentiful seafood once more.

Right: Satellite data from the 2002–03 El Niño shows anomalies in sea surface temperature (SST) and the direction of the wind in the Pacific Ocean. Land is grey. The SST data shows areas that are hotter (orange) and cooler (blue) than normal. Wind direction is shown by the arrows. The El Niño weather cycle is caused by the warm equatorial sea current seen in the Pacific. This is a mild El Niño, where the warmer area is mostly in mid-Pacific, rather than close to the South American coast. A strong El Niño can cause unseasonal storms as well as droughts.

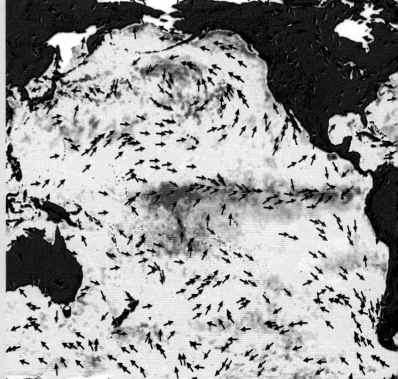

Sea Level Rise

One of the most worrying potential effects of global warming, natural or anthropogenic, is the predicted rise in sea level. Already a number of low-lying island nations, such as Tuvalu, have experienced problems with flooding but the pattern is predicted to continue and to become more widespread.

In a report published in early 2006 Australian researchers led by Dr John Church found that sea levels had risen by 19.5 centimetres/ 8 inches between 1870 and 2004 with most of that having been during the latter 50 years. The Commonwealth Scientific and Industrial Research Organization used tide gauges from around the world to prepare the information on which their conclusions were based. They went on to predict future sea level rises of up to 34 centimetres/ 13 inches during the 21st century if current trends continue. The international organization the Intergovernmental Panel on Climate Change (IPCC) predicted broadly the same conclusions although a worst-case scenario indicated a possible 88 centimetre/35 inch increase during that time.

The cause of the projected increase is the warming of the Earth's atmosphere which will lead to a number of events. One of these is the melting of mountain glaciers and ice sheets such as the Greenland ice sheet. Thermal expansion of the sea water is also predicted and would be a future cause of sea level rise. The West Antarctic ice sheet is of some concern to scientists at present. They estimate that if it melted the sea level could rise by 6 metres/ 20 feet. It is generally believed the sea level was 6 metres/20 feet higher than today during the last inter-glacial period, the time when virtually no permanent ice existed on the Earth's surface.

Numerous glaciers and ice sheets around the globe are now quite obviously retreating or melting.

A complete melting of the Greenland ice sheet would, it is believed, cause a 7-metre/23-feet rise in sea levels but even a partial melt could see a rise of 1 metre/3 feet. A rise of this magnitude would entirely submerge the Maldive Islands and devastate countries like Bangladesh, 50 per cent of which lies below 5 metres/16 feet above sea level. Areas such as the Nile Delta would also become virtually uninhabitable. All of these doomsday predictions would lead to massive population movements with millions of people being displaced and providing national and international governments with, amongst other things, an enormous refugee problem.

Left: Icebergs from the Roseg glacier in a lake in the Roseg valley in the Swiss Alps. The valley used to be covered in ice from the Roseg and Tschierva glaciers. Warming of the climate has led to these glaciers retreating in the last few decades. A moraine (pile of rocky debris) left by the retreat of the Tschierva glacier has blocked the valley and led to the formation of this lake. The Roseg glacier flows on to the lake, into which it calves icebergs.

The Gulf Stream

The Hollywood blockbuster *The Day After Tomorrow*, released in 2004, brought the attention of a wider audience to importance of the Gulf Stream in determining the weather in the northern hemisphere. The film centres on a meteorologist who predicts that the Gulf Stream will shortly shut down and bring a series of catastrophic weather events that will precipitate a new Ice Age.

So, what is the Gulf Stream? It is the strongest ocean current in the northern hemisphere which carries about 135 billion litres/30 billion gallons of water per second. It is a branch of the global thermohaline circulation (THC) that is driven by differences in temperature and salinity. It brings warm water north from the Caribbean to the UK and the rest of northwest Europe. It is responsible for the latter region's temperate climate. Despite being at broadly the same latitude as Newfoundland in northeastern Canada, the UK never suffers from the extreme cold that causes the sea to freeze. The Gulf Stream is responsible for the longer growing season of the coastal parts of western Europe – it is possible to grow semi-tropical plants as far north as Scotland due to the warming effect of the current.

If, however, global temperatures continue to rise then, it is feared, the Greenland Ice Cap will melt. Rainfall will replace snowfall at high latitudes causing potential disruption as it will increase the freshwater and decrease the salinity content of the North Atlantic. This, it is feared, could lead to the complete shut-down of the North Atlantic circulation. A number of computer climate models exist, each predicting a slightly different scenario; nonetheless each suggests that there will be significant disruption to normal weather patterns. If the Gulf Stream shuts down then northwest Europe will become very much colder. However, this would take a lot longer than the film suggests, although very recent research published in 2006 has suggested it may take as little as a few decades. The UK could see temperatures fall by as much as 3°C/5°F or 4°C/7°F which would mean very much colder winters according to the Hadley Centre, a meteorological research institute in the UK specializing in global climate. The difference in temperatures between the 'Little Ice Age' of the 16th and 17th centuries and the medieval warm period was only 1–2°C/2–4°F so 3–4°C/5–7°F is quite significant. The Hadley Centre also says that summers would become shorter and cooler, creating a shorter growing season, and so the implications for agriculture are immense.

Ice core samples from the Polar regions show regular climate changes have occurred in the past with temperature fluctuations of perhaps as much as 5–10°C/9–18°F in as short a time as a decade. Recent research implicates oceanic circulation in many of these changes.

Opposite above: Water flow in the Gulf Stream seen from the shuttle Endeavour. This view, looking roughly southwest, shows the boundary between water in the Gulf Stream and the coastal waters off the eastern USA. Sunglint highlights the difference between the two – the warmer, fast flowing Gulf Stream (lower half) shows complex eddies whereas the coastal waters are calmer.

Above right: Montbretia flowering on the inner Hebridean island of Easdale on Scotland's west coast. The west coast of Scotland lies on the Gulf Stream which brings moist, tropical air up from the Caribbean across the Atlantic. This allows a rich variety of flora and fauna to thrive in what would otherwise be a hostile environment for some species of flowers and plants.

Right: Map showing the course of the Gulf Stream.

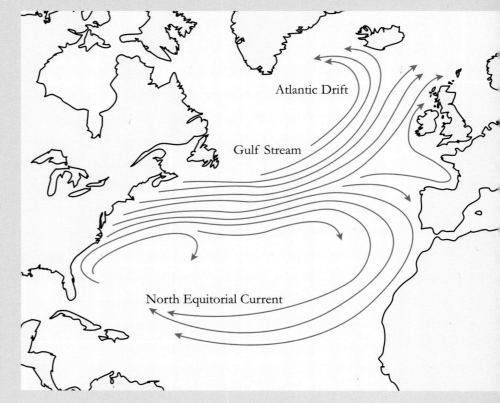

Increased Storms

The summer hurricane season of 2005 was particularly famous for the effects of Hurricane Katrina, which devastated New Orleans and much of the Gulf of Texas in the USA. Inevitably, this and the other powerful hurricanes of that season fed the rumour mill with many suggesting that this was the precursor to a marked increase in the effects of global warming.

According to a 2004 report from the Georgia Institute of Technology the 15-year period 1975–89 experienced 171 severe hurricanes, that is those of category 4 or 5 on the Saffir-Simpson scale. They reported that the figure rose to 269 in the period from 1990 to 2004. The reason for this, they argued, was that the sea surface temperature had increased, a particular manifestation of global warming, and was now about 0.9°C/2°F above the long-term average.

Using 22 computer models of different climate scenarios they deduced that there was an 84 per cent chance that human activity was responsible for at least 66 per cent of sea surface temperature increase in the Pacific Typhoon and Atlantic Hurricane areas.

However, there are year-on-year differences – 2006 was rather quieter than the previous year – so we need to look at more than a few years at a time. It may also be worth considering that today's modern technology is much more sophisticated than in the past. Use of satellite technology to track hurricanes is only a few decades old so making direct and accurate comparisons with events before the 20th century is fraught with difficulties.

What is understood is that hurricane formation over the Atlantic appears to occur in phases which change every few decades. The 1940s and 1950s were active; the 1970s and 1980s were not. The current, more active phase appears to have started in about

Right: Torrential rain, which may accompany a hurricane or other violent storm, often causes flooding and triggers landslides and mudslides. These may then sweep away roads, bridges, railways and sometimes even whole towns.

Below right: Coloured satellite image of Hurricane Floyd (blue, centre left) over the Atlantic. This image was used by the US Air Force Hurricane Hunters team to plan their flight path. The team fly a Hercules plane through hurricanes to gather data. The plane is equipped with instruments to measure meteorological parameters, such as wind speed, humidity and atmospheric pressure. These data are then modelled on computer to help predict the storm's future direction. These models have an accuracy of 70 per cent, which allows authorities to implement damage limitation procedures.

Opposite: Homes and cars near Oklahoma City, USA, destroyed by a force 5 (F5) tornado or twister. Winds in an F5 tornado travel at between 420 and 512 kilometres/ 261 and 318 miles per hour, causing massive devastation.

1995. It is also worth remembering that not all hurricanes hit land and so not all are reported as thoroughly as was Katrina.

In Europe, there was the Great Storm of 1987 over northern France and southern England, and the floods that affected much of Europe in the summer of 2002 – with the flooding of Vienna, Austria being the worst for a century. There are numerous other reports worldwide of local, damaging floods such as these, caused by unusual storms. These events are widely reported – in part because they reflect increasing concern about the frequency and severity of storms, other than those of hurricane force, and their likely association with global warming. It is these less severe storms that may affect more and more people as time passes. We also ought to reflect on the fact that more people now live on flood plains than ever before and so it is inevitable that even small floods will have a greater impact on homes and property.

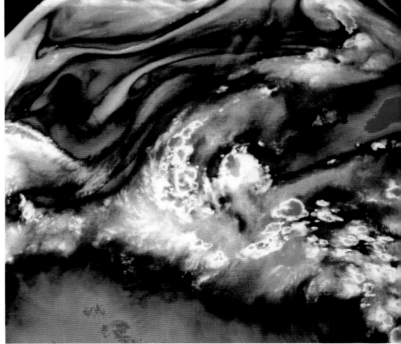

Heatwaves

A total of 740 deaths in Chicago, USA, in the summer of 1995 and almost 15,000 deaths in France in the summer of 2003 were attributed to heat waves of unusual intensity. These are figures which seem alarming but which many climate scientists believe we should come to expect if predictions of increased frequency of such heat waves because of global warming are to be believed.

The United States Center for Atmospheric Research (NCAR) used climate modelling techniques to predict the geography of future heat waves and concluded that Europe and North America were most likely to experience more frequent, longer and more intense heat waves. These were predicted to coincide with an atmospheric circulation intensified by an increase in the amount of greenhouse gases present.

Particularly severe heat waves were predicted to occur in a belt stretching from the Mediterranean Sea and western USA northwards. The Balkans, France and Germany may be increasingly susceptible to heat waves such as those of 2003. The city of Chicago was predicted to experience a 25 per cent increase in such events whereas Paris would experience

up to 31 per cent more, posing problems for the city authorities as they prepare for the future.

Chicago 1995

July 1995 saw temperatures in this city reach a peak of 41°C/106°F but also night-time temperatures frequently as high as 25°C/77°F. Record humidity levels also affected the city making conditions almost unbearable for the resident population. Those who had air conditioning tended to fare the best. Inevitably, though, the people who suffered the most – and therefore contributed most to the death toll – were the poor and elderly who, even if they had air conditioning, could not afford to use it every day or were afraid to open windows for fear of crime. The geographical pattern of deaths was revealing, the poorest areas showing the highest death toll.

Contributing to the problem in a city like Chicago is the 'heat island' effect. Buildings absorb heat during the day and release it slowly during the night. As a result, cities get hot and stay hot under these conditions. A temperature inversion also develops, whereby air temperatures in the atmosphere are cooler at ground level, a reversal of the normal pattern. This means that the relatively cooler, and so denser, air traps pollutants at ground level. They tend to form where there is little wind so the pollutants are not dispersed but persist, adding to the problems.

Chicago was wholly unprepared for these circumstances – there were power failures, ambulance services could not cope and hospitals quickly became overcrowded. The situation is predicted to recur with increasing frequency in the future – perhaps 1995 should serve as a warning for things to come.

Paris 2003 and 2006

Record temperatures were experienced all over Europe during late July and early August 2006. Paris and Berlin reached 35°C/95°F and in Britain the temperature reached 36.5°C/98°F just outside London. Several hundred people are believed to have died as a result of the high day- and night-time temperatures across Europe but this number is low compared to the 35,000 who died during the summer of 2003.

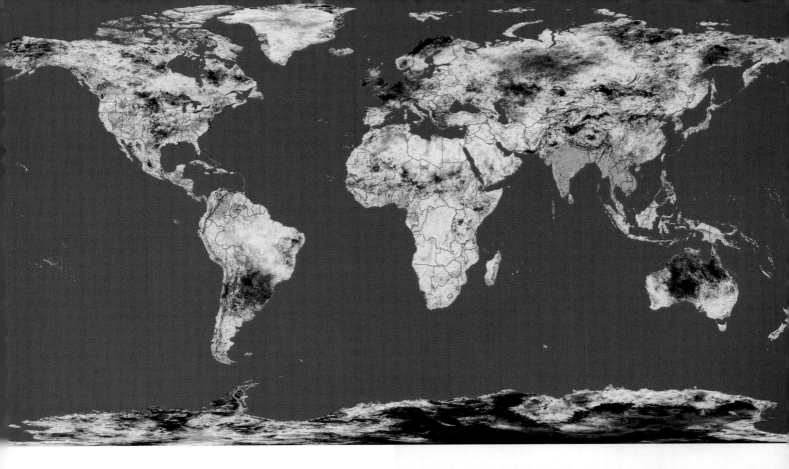

With daytime temperatures reaching 40°C/104°F in a country where air conditioning is as rare as such high temperatures, it is not surprising that France suffered so badly during the heat wave that affected European countries. Scientists at the French National Institute of Health and Medical Research (INSERM) published a report in September of that year stating that the number of deaths had reached 14,802, and many who died were elderly.

The high death toll caused controversy for the medical and emergency services. August is traditionally the month when cities like Paris become virtually deserted as people take their annual summer holiday. No doubt the exceptionally high temperatures that year encouraged others to join the exodus. The health service, however, was accused of being under-staffed at this critical time and those responsible for the care of the elderly were particularly criticized. If the summer of 2003 is typical of the weather to come in Europe then national governments across the continent will have to plan carefully for the future to avoid such a disasters.

The lower death toll in 2006 is likely to be the result of lessons learnt from the disastrous effects of 2003.

Above: Global map showing the increased temperatures compared to previous years, during the heatwave in the Northern Hemisphere in the summer of 2006. The average land surface temperature from 12–19 July 2006 was compared to the average temperature during the same period averaged over the previous six years (2000–2005). Red indicates hotter temperatures and blue indicates cooler temperatures. The extremes (dark blue and dark red) were 10°C/18°F cooler and hotter. Areas with increased summer heat include the USA and north-western Europe. Records for July temperatures were broken in Europe, and crop losses and high electricity demands were reported. Data obtained by the MODIS sensor on NASA's Terra satellite.

Opposite: South Koreans enjoy a swim at Everland amusement park on 10 August 2004 in Yongin, South Korea. The park attracted large numbers of visitors seeking shelter from high temperatures, which soared to 35°C/95°F, the highest level for 10 years in the capital.

Disease

Global warming is predicted to have a significant impact on the future global distribution of diseases. Those currently found only in the Tropics may in future be common in higher latitudes if predicted temperature rises take place.

One disease for which there is currently no preventive vaccine and which affects some 600 million people worldwide is malaria. The malarial mosquito is the *Anopheles gambiae* variety found in the Tropics and the disease is carried by the female alone. When the female mosquito takes a blood meal from an infected person it may take up the plasmodium parasite, the cause of the disease. The mosquito lays its eggs on a stagnant water surface from which the larvae hatch. These will be carriers of the parasite

and can go on to spread the disease further. If warming does take place in higher latitudes then it may be possible for the disease to spread to countries such as those in Europe which currently are malaria-free. Already malaria has been detected in new and higher-elevation areas in Indonesia. The cost to those countries' economies will be immense, to say nothing of the impact on human health.

Mosquitoes carrying other diseases are already spreading as climate shifts allow them to survive in formerly inhospitable areas. Mosquitoes that can carry dengue fever viruses were previously limited to land up to 1000 metres/3300 feet but recently appeared at 2200 metres/7200 feet in the Andes Mountains of Colombia.

Those areas currently affected by such diseases tend to be amongst the poorest countries of the world.

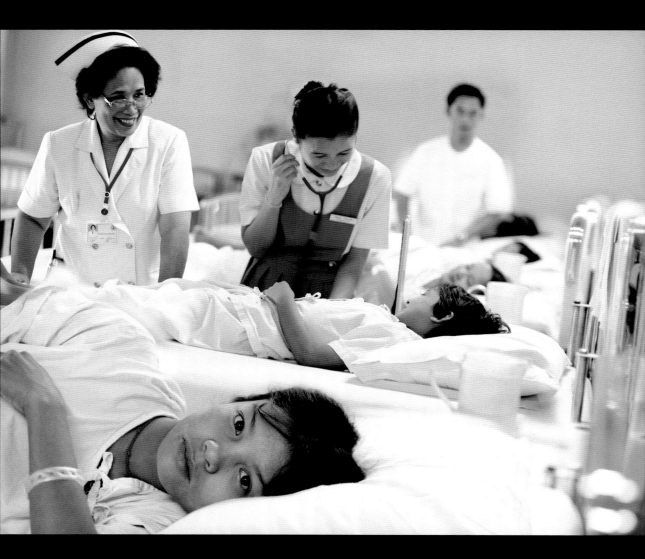

Population Migration

About 12,700 years ago, there was a cooling of at least 15°C/27°F in Greenland, and substantial change throughout the North Atlantic region as well, lasting about 1300 years. This event had an enormous effect on the ocean and land surrounding Europe – icebergs were found as far south as the coast of Portugal. Today its impact would be more severe; Europe's population is now much greater in size and in density.

The more recent periods of cooling, such as the time of the 'Little Ice Age' from the 16th to the 18th century, appear to be intimately connected with other changes, civil unrest, inhabitability of once desirable land, and even the demise of certain populations.

With over 400 million people living in drier, sub-tropical, often over-populated and economically poor regions today, climate change and its associated effects pose a severe risk to the political, economic and social stability of many nations. In poor countries where there is a lack of resources and capabilities to adapt quickly to more severe conditions, the problem is very likely to be exacerbated. For some, mass emigration may be the result as desperate people look for a better life in regions such as the USA that have the resources to adapt. One such country is Bangladesh. One prediction for the country by the mid-21st century is for persistent typhoons and a higher sea level storm surges that will cause significant coastal erosion and make much of the country uninhabitable. As sea levels rise there will be contamination of fresh drinking water supplies, leading to a humanitarian crisis that is predicted to lead to mass emigration. Tension between Bangladesh and China and India, all of which may struggle to cope, will increase as the migrants head for those countries.

It has been suggested that the continent to be hardest hit by climate change, however, is Europe. A temperature fall of 3°C/6°F in less than a decade has been predicted. Even more dramatic falls along the northwestern coasts are predicted making the area drier, windier and much more like Siberia. Mass emigration from Scandinavia and other northern European countries may take place as people search for a warmer climate. Southern Europe may be squeezed by both the emigrating northern Europeans and those leaving the hard-hit parts of Africa. Emigration from the latter continent has already been a common feature – and controversial issue – in Europe for many decades. How well the continent will be able to cope is difficult to predict but it is unlikely to be easy.

Above: The Inuit are the descendants of the Thule culture and are a nomadic people who emerged from western Alaska around the year 1000 and spread eastwards across the Arctic. By 1300 they had settled in west Greenland, and finally moved into east Greenland in the following century.

Opposite: Many scientists fear that climate change will encourage the spread of disease.

WEATHER AROUND THE WORLD

NORTH AMERICA

Canada

The Maritime Provinces

The climate of Canada's eastern coast, which includes Newfoundland, Labrador, New Brunswick, Nova Scotia and Prince Edward Island, is chiefly moderated by the maritime influence of the Atlantic Ocean. This tends to result in less severe winter temperatures and cooler summer temperatures than are typically experienced in much of the interior. However, the weather can also be highly variable throughout the year, largely on account of the numerous cyclonic depressions that move across the area from the Great Lakes region, and cloud and precipitation are a year-round phenomenon, with heavy snowfalls being a common feature in winter. Fog is also a major characteristic of the Maritime Provinces, particularly around Newfoundland and the Gulf of St Lawrence, and in fact St. John's in

Newfoundland is Canada's foggiest city. Sea fog can be especially persistent in summer, and results mainly from the meeting of the cold waters of the Labrador Current and those of the warm Gulf Stream. Icebergs drifting south in the Labrador Current pose a further hazard for shipping in summer. Winters are often relatively mild in the region, with average temperatures ranging from about −7 to 0°C/19–32°F in February, whilst in summer, the average range for July is between about 11–20°C/52–69°F. Despite this, seasonal lows of −51°C/−60°F and highs of 42°C/107°F have been recorded in Newfoundland. Annual precipitation in the Maritime Provinces averages about 1150–1500 millimetres/45–60 inches.

Southern and Central Ontario and Québec

This region has a generally temperate, continental climate, with fairly mild or warm summers and cold winters, but as in the Maritime Provinces of eastern Canada, the weather can be highly changeable throughout the year. Relatively high precipitation is distributed across the seasons, and conditions are subject to the occurrence and movements of frequent cyclonic storms.

The south of the area, which is bordered by the United States and the Great Lakes, is the most southerly, highly developed and densely populated part of the country, and the latitude and presence of the lakes also combine to make it one of the warmest. However, whilst winters may be typically less severe than in parts of northern and western Canada, they can be long and cold even here, and with the exception of parts of the Rockies, are subject to some of the heaviest snowfalls, with several feet falling annually, and snow frequently covers the ground from December to March in many places. Average winter temperatures in Ottawa and Québec range from about −15 to −6°C/5 to 21°F, but tend to be warmer in Toronto, which lies on the edge of Lake Ontario.

Summers are often warm, with several hours of sunshine, and July temperatures will usually average between about 15–26°C/59–79°F for much of the region.

The Prairie Provinces

With the exception of the far north of the region, and those mountainous parts occupied by the Rockies in Alberta, the Prairie Provinces of Manitoba, Saskatchewan and Alberta experience a decidedly continental climate, with distinctly contrasting winter and summer conditions. The winters tend to be long and are frequently severe, with the lowest temperatures often failing to rise greatly above those experienced much further north in the sub-Arctic. The summers, although relatively short, are typically warm and mild, or may even be quite hot. Additionally, whilst the Great Plains to the south are mainly semi-arid, and the Canadian Prairie Provinces are also relatively dry, receiving around 250–1000 millimetres/10–40 inches of precipitation annually, spectacular thunderstorms may take place in summer, when most rainfall occurs in the region, and in Alberta hailstorms are also common. In the winter, meanwhile, snowfalls and severe blizzards are also experienced, although the snowfall is typically less extreme than in eastern Canada for example, and snow that has settled often thaws or is cleared by wind before falling again.

The region is perhaps most notable for the long hours of sunshine received throughout the year, which average between about three to five hours in winter, and eight to ten hours in summer, and the city of Estevan in Saskatchewan is the sunniest in Canada, with about 2540 hours of sunshine each year, whilst Alberta is also sometimes known as the 'Sunshine Province'.

The summer temperature in these provinces tends to average about 19°C/66°F, and the winter temperature between about −13–0°C/8–32°F, although in Calgary, Alberta, the winter average is often raised by the mild Chinook winds, which are warmed as they cross the Rockies from the Pacific.

British Columbia

The climate of British Columbia is generally relatively mild, but varies a great deal according to topography; much of the area is mountainous, with high plateaus and deep valleys, whilst there are areas which extend into the prairies, and those which lie along the coast of the Pacific Ocean.

Along the Pacific coast and in some of the valleys inland, both the summers and winters tend to be mild, and the temperature range is the narrowest

Left: Much of Canada is uninhabited because the severity of the climate and sea temperatures can also be very low. In summer, icebergs are a hazard to shipping on the east coast.

Above: Steady winds turn the test windmills at the Alberta renewable energy test site at Crowsnest Pass in the Canadian Rockies towards the prairies.

Opposite: A beautiful display of autumn colour at Larch Valley, Alberta.

in Canada, ranging from an average of about 4°C/39°F in January, to around 16°C/61°F in July. The coast also has the warmest winter temperatures in Canada, and receives a lot of precipitation, which ranges from about 1500–3000 millimetres/60–120 inches, and increases at higher elevations in the Coast Mountains, where much of it may fall as snow. In some of the valleys, however, annual rainfall may be as low as 375 millimetres/15 inches. Along the coast, cloud cover and fog are also common, reducing winter sunshine to perhaps 2–3 hours per day, although during the warm summer, between eight and ten hours may be experienced. Inland, particularly on the prairies, the climate is typically more continental, with winters being longer and more severe, and summers being warmer, with a much wider range of temperatures across the seasons, and even between day and night. However, overall, British Columbia's climate is often regarded as amongst the most favourable of the Canadian provinces.

Northern Canada

The climate of northern Canada, including the Yukon, the Northwest Territories, Nunavut, and the northernmost parts of British Columbia, the Prairie Provinces, Ontario and Québec, is the harshest in the country, and may be described as primarily Arctic or sub-Arctic, depending upon latitude. Winters are typically very long and cold; lasting up to a staggering 11 months in places, and along the Arctic Circle, sub-zero temperatures may endure for around seven months of the year, with average January temperatures ranging between –35 to –20°C/–31 to –4°F. Conversely, the summers in northern Canada are usually short and sunny, and are also generally quite mild, generally rising to a July average of around 10–14°C/50–57°F in parts of the Yukon, Northwest Territories, Nunavut, and slightly higher further south. However, at Alert, the world's most northerly settlement, average temperatures may rise above freezing only briefly at the height of summer. Overall, precipitation is relatively low throughout the year, with light snowfall in winter and light rains in summer.

Much of the northern coast of Canada may remain permanently frozen, and even Hudson Bay is frozen for much of the year. On land permafrost may extend up to hundreds of feet/metres beneath the ground in places, and extends in a belt from the Yukon to the south of Hudson Bay and across to Labrador. Seasonal variations in daylight hours also play an important part in the climate of the region, ranging from as much as 24 hours of daylight in June to 24 hours of darkness in December, whilst windchill can further dramatically reduce temperatures, particularly in winter.

USA

Alaska

Being geographically separated from the contiguous states of the USA, and bordering the Canadian Yukon, Alaska's climate more closely resembles that of northern Canada than anywhere else in the USA. However, its climate also varies from region to region, on account of latitude, topography and maritime influences.

Northern Alaska has an Arctic or sub-Arctic climate, and the northern Arctic Ocean coast may be frozen for much of the year. There is also permanent ice and snow at higher elevations, and permafrost in low-lying areas. Winters are long and severe; in the northernmost town of Barrow, for example, temperatures regularly fall to $-31°C/-24°F$ in February. Summers, although short, can be surprisingly warm, with temperatures of up to about $21°C/70°F$ in places. Annual snowfall is moderate, whilst rainfall is low – in fact Barrow experiences desert conditions, since it has less than 100 millimetres/4 inches of rain per year.

Alaska's interior also experiences very severe winters, but the climate is more continental, with great temperature extremes between summer and winter. In winter it may be as cold as $-51°C/-60°F$ for weeks at a time, but in summer temperatures may climb to above $32°C/90°F$.

In the south, along the coasts of the Bering Sea and Gulf of Alaska, conditions are different again. The winters are typically far milder, but are often accompanied by heavy snowfalls, whilst the summers tend to be rather cooler than they are further inland. Annual precipitation tends to be highest in the southeast, and Little Port Walter in this area is the wettest place in Alaska, with an average of 5588 millimetres/220 inches.

The Northwestern USA

The northwestern USA includes the states of Washington, Oregon, Idaho, Montana and Wyoming, stretching from the Pacific coast to the Great Plains. However, a broad distinction can be made in this region between the climate of the Pacific Northwest, which is taken to include Washington and Oregon, where mild maritime conditions prevail, and the more continental climate that is experienced further inland across the states of Idaho, Montana and Wyoming.

Above: Giant redwood trees are found near the coastline in Oregon and California.

Opposite: Mount McKinley in Alaska is snow-covered all year round.

Along the coast, the summers tend to be cool, and the winters mild, with temperatures being moderated throughout the year by the prevailing westerly winds from the Pacific Ocean. In fact, the area has the smallest annual temperature range in the USA, averaging between about 15–20°C/59–68°F in summer, and 4–10°C/40–50°F in winter. The area is, however, amongst the wettest in the USA, with moist air blown from the Pacific depositing its precipitation as it rises up the west slopes of the Rockies, Cascades and Olympic mountain ranges and cools. In some places annual rainfall may be as high as 3560 millimetres/142 inches, most of which occurs in winter.

Above: The monarch butterfly takes advantage of the diversity of the American climate, wintering in the southern states and returning north in spring.

Opposite above: Tornadoes, as well as thunderstorms and hailstorms. may occur during the summer months in the Midwest states.

Opposite below: Inhabitants of Chicago try to free their cars covered in a thick layer of snow.

In the northwestern interior, moderating maritime influences are still experienced in parts of Idaho, Montana and Wyoming, particularly in the west, where the mountain ranges bring increased precipitation. The eastern areas are typically more arid and continental, with lower rainfall, cold winters and hot summers, although in winter, Chinook winds can make for milder conditions. Average temperatures may range from as low as –12 to –6°C/11 to 22°F in winter, with summer highs of 29–35°C/84–5°F, whilst the annual precipitation may be as low as 127– 305 millimetres/5–12 inches on the plains, and as high as 508 millimetres/20 inches or more in the mountains – which may also receive high snowfalls. Severe hailstorms are also quite a common feature of the region.

The Great Plains

The states of Minnesota, North Dakota, South Dakota, Nebraska and Kansas, constitute the High Plains; a region with a generally semi-arid climate, where annual precipitation is low and unpredictable, and the seasonal range of temperatures is extreme. Winters on the plains can be severe, particularly in the north, such as in Minnesota and North Dakota, where temperatures may regularly fall to as low as –13°C/9°F or even –18°C/–1°F. They tend to be somewhat milder further south, with the coldest temperatures in parts of Kansas dropping to about –8 to 3°C/17 to 37°F. In the northwest, meanwhile, warm Chinook winds may help to moderate winter temperatures, and what little precipitation occurs in winter tends to fall mainly as snow. The southern parts of the region are also generally warmer in summer, with average highs of 32°C/89°F recorded in parts of Kansas, compared to 28°C/83°F in North Dakota, and 24°C/76°F in Minnesota. However, the highest recorded temperature for the region was at Steele in North Dakota, where it reached 49°C/121°F in July 1936. As in winter, precipitation is also relatively low in summer, and frequent droughts occur in many areas across the High Plains, although violent thunderstorms may occur, bringing much of the annual rainfall in perhaps just a matter of days. Overall, rainfall for the region averages less than 510 millimetres/20 inches and the south and northeast tend to be the wettest areas, with precipitation decreasing further west.

area is −48°C/−54°F, measured in Wisconsin in January 1922. However, winter temperatures in the state usually average between about −13 to −3°C/9 to 27°F. Further south, such as in parts of Ohio and Missouri for example, January temperatures may often exceed 4°C/39°F. Summers are also warmer in the south of the region, while the east tends to be wetter, particularly in winter. Average precipitation for the Midwest and Ohio Valley region ranges from about 500–1000 millimetres/20–40 inches per year.

New England and the Mid-Atlantic

Despite the region's proximity to the Atlantic Ocean, the New England and Mid-Atlantic area has an essentially humid continental, rather than maritime climate, which is influenced more greatly by air masses moving in from the west than by those from the Atlantic. The area can also be subdivided into the northern states of Maine, New Hampshire, Vermont and upper New York, which experience longer, colder winters and warm or cool summers, and the area which includes southern New York, Massachusetts,

The Midwest and Ohio Valley

The climate of this region, which encompasses the states of Iowa, Missouri, Wisconsin, Illinois, Michigan, Indiana, Ohio and Kentucky, is essentially humid and continental. It is transitional between the humid subtropical climate of the states to the south and southeast, from which warm air moves northwards, and the sub-Arctic climate of Canada to the north, from which cold air descends.

The area is characterized by four distinct seasons, with cold winters, warm springs, warm or hot summers – which sometimes bring droughts – and fairly mild, wet autumns. The range of average temperatures experienced throughout the year is generally wide, often descending to below freezing in winter, when there may be prolonged snow on the ground, and rising to about 18–24°C/65–75°F in summer. Summers also tend to be humid, and thunderstorms, hailstorms and tornados may occur at this time of the year.

Winters within this region tend to be more severe further north around the Great Lakes and the border with Canada, where snowfalls may also be heavy, and the coldest temperature recorded for the

Rhode Island, Connecticut, Pennsylvania, New Jersey, Delaware, Maryland, the District of Columbia, West Virginia and Virginia, where the summers are typically hotter and often also wetter.

In the northern part of the region, there may be prolonged periods of snow in winter, and during January, temperatures may average around -15°C/5°F, whilst further south, in Richmond, Virginia for example, temperatures for the same period average between about –2 to 8°C/28 to 47°F. In winter, 'Nor'easter' storm systems may also travel up the coast, bringing high precipitation to coastal areas and often heavy snow inland.

In summer, temperatures in the north at Portland, Maine, range from about 14–26°C/58–79°F, whilst in the south at Richmond, highs of 31°C/88°F may regularly be achieved, and heat waves can witness temperatures in excess of 38°C/100°F. Precipitation is relatively high throughout the year, averaging about 800–1000 millimetres/32–40 inches, and it tends to occur with thunderstorms in summer when the air is warm and humid. Aside from stormy periods, summers tend to be sunny throughout the region, although fog may also occur, particularly in coastal parts.

Above: Sunset over Alabama's Gulf Coast, which can be hit by hurricanes.

Below left: Fall colour in New York.

Opposite: At Yosemite National Park elevation plays a major part in determining the weather.

The Southeastern USA

The southeastern USA, which includes the states of Tennessee, North Carolina, South Carolina, Georgia, Alabama and Florida, has a largely humid subtropical climate, characterized by long, hot, humid summers with a great deal of sunshine, and cool, mild, but often cloudy winters. Average winter temperatures range from about 4–10°C/40–50°F overall for the region, but may be lower in the southern Appalachian mountains, which extend into parts of Tennessee and Alabama, and much higher in Florida, which is almost tropical, and where the average January temperature ranges between 10–22°C/50–72°F. Frost and snow are rare phenomena in southern Georgia and Florida, with most snow that affects the southeast falling much further north and farther inland. However, large storm systems known as 'Nor'easters' may bring cold rains to much of the coast in winter.

A great deal of precipitation is also received in summer, much of it brought by thunderstorms. These increase in frequency southwards, so that Florida experiences more than any other US state, with often over 100 storms annually. Atlantic Hurricanes also occur in the region, typically between July and October, and these may bring not only heavy rains, but devastation along the Gulf coast. Despite such storms, the Southeastern US benefits from relatively long hours of sunshine throughout the year, and in summer, temperatures average about 26°C/80°F. In the south, particularly in Florida, maritime influences tend to moderate the temperature, so there is a relatively small annual range.

The average annual precipitation ranges from about 760 millimetres/30 inches in the west, to over 1,500 millimetres/60 inches in the south.

The Southwest

Much of the southwest region, which incorporates Nevada, Utah, Colorado, Arizona, New Mexico and interior California, is extremely arid, with desert or semi-desert conditions, whilst much of the coast of California has a Mediterranean-type climate.

Away from the coast, desert conditions in the southwest interior prevail largely on account of the 'rain-shadow' effect, which is produced by the mountains of the Pacific Cordillera and Sierra Nevada ranges. These act as a barrier to the moisture-laden air masses that are drawn over them from the west, which causes precipitation to occur on the windward side, whilst the air is warmed again as it moves down the leeward side. The lack of moisture in the air produces very clear desert skies, which in turn allow temperatures to increase significantly in these areas during the daytime, and also to cool very rapidly and dramatically at night.

Precipitation for the region averages less than 250 millimetres/10 inches per year, although there are several areas that may receive as little as 80–125 millimetres/3–5 inches, and mountainous areas in New Mexico and northern Arizona that may receive as much as 760 millimetres/30 inches. The summers are long and extremely hot in the desert, whilst the winters are short and mild. Average temperatures range from between about 29–35°C/85–95°F in summer, to between about 18 and 24°C/65–75°F in the winter. However, the highest temperature recorded in Death Valley California is 57°C/134°F, and the coldest, –9°C/16°F.

The climate of California's northern coast is similar to that of the Pacific Northwest, but further south, it becomes increasingly warm and dry in summer, and mild, with moderate rainfall in winter. However, San Francisco experiences a microclimate, whereby summers are cooled by the recurrent sea fogs, with average July highs of just 22°C/71°F, compared with 27°C/81°F in Los Angeles.

Hawaii

The Hawaiian island chain comprises the only USA state outside of the North American continent, and it is also the only state that possesses a tropical climate. However, due to moderating maritime influences and topography, its climate is often regarded as being closer to subtropical, and both temperatures and humidity are typically lower than is common for tropical islands of a similar latitude. In low-lying areas, sea breezes contribute to these moderating effects and inland the islands are hilly or mountainous, and altitude plays a significant important role.

Temperatures show little variation throughout the year, ranging in Honolulu for example, from about 23°C/73°F in January, to around 27°C/80°F in July. Temperatures may fall below 18°C/65°F in some places in winter and at high altitudes there may even be snowfalls during the winter months.

In addition to having a pronounced effect on temperature, the prevailing northeast trade winds are also responsible for bringing heavy precipitation to the exposed northeastern parts of the islands, and in fact, parts of Hawaii are considered the wettest in the world. A record 1878 millimetres/739 inches have been recorded in a single year at Kukui, Maui, whilst 300 millimetres/12 inches have fallen in an hour at Kauai, Hawaii, and Mount Waialeale on Kauai averages 11,680 millimetres/460 inches of precipitation annually. Remarkably however, due to the effects of 'rain-shadow', some of the areas on the leeward sides of the wet hills and mountains, particularly areas in the southwest, may receive just 640 millimetres/25 inches.

Between May and November, tropical storms may strike Hawaii, but this tends to be a relatively rare occurrence, with rather less severe storms than in many areas that are affected by hurricanes and typhoons.

Left: Palm trees on Waikiki Beach, Oahu, Hawaii.

Opposite: Mayan structure called El Castillo/The Castle on the Caribbean coast, at Tulum Archaeological Site, Mexico.

Mexico and Central America

Mexico

Mexico is effectively divided into a northern temperate and southern tropical region by the Tropic of Cancer. However, with mountains and high plateaus occupying much of the country, the topography has a marked effect on the climate, and the surrounding waters of the Pacific Ocean and Gulf of Mexico are also of significant influence. Parts of the northwestern coast around Tijuana for example, experience a Mediterranean-type climate, with mild conditions, significant sea fog and wet winters. Conditions elsewhere range from hot and humid in many coastal areas, particularly in the south, to relatively mild throughout much of the Central Plateau, where it is seemingly almost always spring-like. However, seasonal variations also occur, and in the north the climate tends to be more continental, with warmer summers and colder winters.

Temperatures in the south, including the Yucatan Peninsula, average between about 24–28°C/75–82°F throughout the year, whilst further north the range is from about 20–24°C/68–75°F in low-lying areas, and between about 8–12°C/46–54°F at higher elevations.

Precipitation also varies widely, both seasonally and according to location, with arid conditions being experienced in much of the north and the Central Plateau, where rainfall averages between about 300–600 millimetres/12–24 inches per year, whilst parts of the southeast may receive in excess of 2000 millimetres/80 inches annually, with most rainfall occurring between June and October, in what is essentially the rainy season. Snowfalls may also occur in the north in winter, and hurricanes are a potential threat along both coastlines from about June until late November.

Central America

The dominant climate types throughout Central America are humid tropical and subtropical, particularly in the low lying coastal areas, whilst at higher elevations a more temperate or even alpine climate may be experienced. In some coastal areas, such as in Belize, however, the prevailing northeast trade winds moderate the temperature and it may be somewhat warmer inland. Temperatures across the region tend to remain fairly high throughout the year, averaging between about 16–32°C/61–90°F in January and July respectively, but at higher altitudes, such as in the mountains of Guatemala, Costa Rica and Panama, the average range is from about 10–27°C/ 50–81°F, being coldest at elevations of over 1524 metres/5,000 feet.

Relative humidity averages around 80 per cent in many areas and most of Central America experiences just two seasons, a wet or rainy season, and a dry season. The rainy season occurs from May to October in Belize, Guatemala, El Salvador and Honduras, and from May until November in Nicaragua, May to December in Costa Rica and May to January in Panama. Some areas however, experience the 'canicula' or 'little summer' in July and August, during which there may be prolonged dry spells. In terms of precipitation, with the exception of parts of El Salvador's interior, which are somewhat arid, much of Central America experiences relatively high annual rainfall, ranging from about 838 millimetres/33 inches in southern Honduras to 3810 millimetres/150 inches in Belize and up to 6350 millimetres/250 inches along the Mosquito Coast in Nicaragua. Much of the rain results from heavy thunderstorms, and it is not unknown for hurricanes to make landfall in the region.

The Caribbean and Atlantic Islands

The climate throughout the Caribbean Islands and across Bermuda in the Atlantic Ocean is generally subtropical, with relatively high annual temperatures with little seasonal variation, and extremes of temperature tend to be moderated by the prevailing trade winds and maritime influences. However, in parts of Jamaica and Haiti, and further south, across Dominica, St Lucia, St Vincent, the Grenadines, Barbados and Trinidad and Tobago, the climate is close to tropical, whilst at higher elevations temperatures may decrease quite significantly, resulting in more temperate conditions. Throughout most of the region there is a dry season from about December to March and a rainy season from about May or June through

to October or November. In some places however, there are two rainy seasons. In Jamaica, for example, these occur from May to June and from September to November, whilst in Haiti they occur from April to June and from October to November. Storms and torrential downpours are common during rainy the seasons, and hurricanes can be a frequent and sometimes devastating occurrence from about August to October or November. Despite the small seasonal range between temperatures, it tends to be warmest during the hurricane season and coolest in the period that follows. Average temperatures range from a winter minimum of about 18°C/64°F to summer highs of about 35°C/95°F. Average annual precipitation is highly varied, ranging from just 533 millimetres/ 21 inches in the Turks and Caicos Islands, and 650 millimetres/26 inches around Guantánamo Bay in Cuba, to around 3800 millimetres/150 inches on the mountain peaks of St Lucia, and as high as 5000 millimetres/200 inches on Jamaica's northeast coast and in the Blue Mountains.

Left: Most of the Caribbean islands have a subtropical climate with little seasonal variation. The temperatures are moderated by the surrounding seas and the prevailing trade winds.

Above and right: In moist tropical rainforests it rains nearly every day and is hot and humid throughout the year. The amount of water vapour carried in the air is high because much of the evaporation cannot rise above the tree canopy to be blow away, so it falls again as rain.

in climate. In northern parts of the continent, which lie close to and north of the equator – including Colombia, Venezuela, Guyana, Suriname, and French Guiana – the climate is hot, humid and tropical, with a rainy season that typically coincides with the Northern Hemisphere summer, between about May and August. In some areas, a second wet season occurs between November and January, and others see almost constant rain. In areas extending into the Amazon, average annual precipitation may be as high as 3000 millimetres/118 inches or more, although the northern coasts of Columbia and Venezuela are surprisingly arid, and both there and in parts of north-eastern Brazil, droughts can be severe during the dry season. There is little variation

the region tend to be somewhat cooler, and conditions along the coasts of Guyana and Suriname are moderated by the northeast trade winds.

Central South America

Much of the region between the Equator and the Tropic of Capricorn is occupied by Brazil and the vast tropical Amazon basin, where the wet season occurs between about November and April, with about 2000 millimetres/80 inches of rain falling during the Southern Hemisphere summer, followed by a dry season from about May to October. However, in Ecuador and the mountains of Peru to the west, the rainy season generally lasts from December until May, whilst much of coastal Peru,

south into northern Chile is effectively desert, receiving virtually no rain throughout the year. Average temperatures here range from about 17–21°C/ 64–70°F in winter, to around 23–27°C/ 73–81°F in summer, becoming cooler at higher elevations, where temperate and alpine climates may be experienced. In the Amazon, temperatures average about 29 to 33°C/84–91°F, with little seasonal variation. In the eastern highlands and southeast of Brazil meanwhile, cool fronts can push northward from Argentina, bringing cloud and winter rains, and lowering temperatures to 10°C/50°F or lower at times – although highs of 32°C/90°F can be experienced in the winter months, and in some areas temperatures may exceed 38°C/100°F in summer.

Towards the southwest of the region, Bolivia's climate ranges from hot and humid in the rainforests and grasslands of the east, to arid in the high plateaus of the altiplano, where the temperature range can be extreme.

The South

Below the Tropic of Capricorn, the climate becomes increasingly temperate through much of Argentina, Chile and Uruguay, and the region known as Patagonia, to the cold maritime climate of the Falklands, and the frigid, sub-Antarctic conditions experienced at Argentina's southern tip. However, further north in Paraguay, the climate ranges from warm, humid and subtropical in the north, to temperate and mild in the south. Rainfall varies accordingly, from about 1520 millimetres/60 inches to 760 millimetres/30 whilst temperatures range between about 16°C/60°F in winter, up to about 38°C/100°F in summer.

In linear Chile, the Pacific Ocean and the Andes are the major influences, but the climate ranges from the incredibly arid Atacama Desert in the north, through a central Mediterranean climate with warm, dry summers and mild, wet winters, to the south, where high winds, known as the 'Roaring Forties' and 'Screaming Fifties', and precipitation in excess of 4,060 millimetres/160 inches are major characteristics.

In Uruguay and central Argentina, summers tend to be warm and humid, with average temperatures of about 25°C/77°F, whilst winters are generally mild, with average temperatures of about 15°C/59°F. Rainfall tends to be distributed fairly evenly throughout the year, with Uruguay receiving about 1050 millimetres/ 41 inches, but this decreases from east to west. Much of Argentine Patagonia is also very dry by comparison, being sheltered by the Chilean Andes, but precipitation can occur throughout the year, and cloud is common in both winter and summer. Temperatures in the region tend to be cool, regularly dropping below freezing in winter, particularly in the windswept far south, but conversely in summer, warm winds may descend from the mountains.

Opposite above: The Valley of Death in Atacama Desert in the north of Chile is one of the most arid areas in this part of the world, receiving virtually no rain in a typical year.

Opposite below: Machu Picchu in Peru, shrouded in morning fog. This area of Peru has an almost sub-tropical climate, although other areas of the country are very dry, with little rainfall.

Above: The Chilean Andes shelter large areas of Argentina, so the climate is quite dry across the prairie although cloud is common. Temperatures are generally cool here.

EUROPE

Northern Europe

Denmark

The climate of Denmark is regarded as temperate, but is also subject to maritime influences. The summers tend to be mild, warm and sunny, and the winters are cold, cloudy and often rainy – although most precipitation occurs from late summer into early autumn. In the autumn and winter, particularly, strong Atlantic winds can also dramatically lower the temperature.

Temperatures average from around –3 to 2°C/27 to 36°F in the coldest month, February, to between about 14 to 22°C/57 to 72°F in July, the warmest. The average annual precipitation is about 600 millimetres/24 inches. Around 10 per cent of this is usually snowfall, which occurs mainly between January and March.

Estonia

Generally, Estonia experiences cold winters and cool summers, with the interior of the country having a continental climate, where the temperature range may be most extreme. The coastal areas along the Baltic Sea, however, are favoured by a maritime climate, which provides relatively mild winters.

Annual precipitation ranges from about 500 to 710 millimetres/19–28 inches, with most occurring in late summer, whilst temperatures range from about –7 to –3°C/20 to 27°F in January, to between 15 and 18°C/60 to 65°F in July.

Finland

Despite its northerly position, Finland's climate is moderated by the waters of the Gulf of Bothnia, Gulf of Finland, the Baltic Sea, and by the North Atlantic Current, as well as its several thousand lakes. However, its climate may be described as cold temperate. Winters are generally cold and long, particularly in the north, where snow-cover may persist for much of the year, whilst summers are cool, with the occasional heat wave.

Average temperatures range from about –9°C/16°F in February, to as high as 23°C/72°F in July, with an annual average precipitation of around 460 millimetres/18 inches in the north, to about 700 millimetres/28 inches in the south, including snowfall.

Greenland

Although geographically part of the American contintent, polically Greenland is associated with Denmark. Over 80 per cent of the land is permanently frozen ice cap, and the climate may be described as polar or arctic in the north, and sub-arctic in the south, particularly in coastal areas where the maritime climate has a moderating effect. Winters are long and cold, and summers are short and cool throughout.

In the north, temperatures range from about –23°C/9°F in February, to 5°C/41°F in July, while in the south the range is closer to between –8 to 10°C/18 to 50°F. Annual precipitation averages around 1219 millimetres/48 inches.

Opposite above: Relatively high rainfall, particularly in the west, means that the Ireland is sometimes known as the 'emerald isle'.

Below: Icebergs from the Vatnajokull Glacier rest in a glacial lagoon in Jokulsarlon, Iceland.

Iceland

Despite its name and northerly position, Iceland's climate is relatively moderate, as it is warmed by the currents of the North Atlantic Drift and the Gulf Stream. However, snow may cover most of the ground from November to April, and it is generally both very windy and cloudy for much of the year. Northern and eastern parts also tend to be chilled by drift ice and polar currents.

In the north, annual precipitation averages around 510 millimetres/20 inches, whilst in the south it ranges from about 1270–2030 millimetres /50–80 inches. The average temperature in the capital Reykjavik ranges from about –1°C/31°F in January, to around 11°C/52°F in July.

Ireland

Ireland's climate is temperate, but modified by maritime influences such as the North Atlantic Current and prevailing Atlantic winds. Conditions are generally mild and humid, but may be highly changeable. Rain is common throughout the country at any time of the year, but most falls in winter, and in the west, whilst low-lying areas in the east typically experience drier conditions.

Annual precipitation averages between 750–1250 millimetres/30–50 inches across Ireland, and temperatures average between 4–7°C/39–45°F in winter, and between 14–16°C/57–61°F in summer; around 14°C/25°F higher and 4°C/7°F lower respectively, than many locations the same latitude.

Latvia

The Latvian climate is generally temperate, although the Gulf of Riga and the Baltic Sea bring a maritime influence to coastal areas. The western parts of the country tend to experience cool summers and relatively mild winters, whilst more extreme variations are experienced in the interior, where summers may be hot and winters more severe with a prolonged ground-covering of snow.

Temperatures average around 20°C/68°F in July, falling to an average of around –1°C/30°F in January, whilst precipitation averages between 560–790 millimetres/21–31 inches per year.

Lithuania

Lithuania's climate is temperate and transitional between the Russian continental climate, and the more moderate maritime climates experienced in Western Europe. Along the Baltic coast, winters tend to be mild and summers cool, whilst inland in the interior regions, more dramatic variations between summer and winter may be experienced.

Temperatures average around 17°C/63°F in July, and –4°C/25°F in January, whilst annual precipitation averages between 850 millimetres/ 34 inches in upland areas near the coast, to about 600 millimetres/24 inches in central low-lying regions.

range is −3 to 18°C/27 to 64°F over the same period. Annual precipitation averages 535 millimetres/21 inches and is concentrated in the southwest.

United Kingdom

England
England's climate is temperate, and as with the rest of the UK. It is subjected to moderating maritime effects, particularly those of the Gulf Stream, which results in more moderate conditions than those experienced at Newfoundland and Labrador in Canada, for example, which are at a similar latitude. Summers tend to be warm, and winters relatively mild, with most rainfall typically occurring in autumn. Eastern parts tend to be drier than much of the west, and the greatest snowfalls are concentrated in the north.

The annual precipitation averages around 760 millimetres/30 inches throughout the country, and average annual temperatures range from about 9°C/48°F in the north, to 11°C/52°F in the south, although London has its own microclimate and is typically slightly warmer than the surrounding areas.

Northern Ireland
Northern Ireland shares the moist, temperate climate that prevails over the whole of Ireland, and is subject to frequent rain, but extremes of weather are typically mitigated by the prevailing Gulf Stream winds and currents, so that summers are generally cooler, and winters milder than elsewhere in the UK. However, Belfast and the east coast often experience cold winds in winter.

The average annual temperature ranges from about 4°C/39°F in January, to 14°C/57°F in July, with an average annual precipitation of around 1016 millimetres/40 inches in mountainous parts and about 760 millimetre/30 inches elsewhere.

Scotland
As with elsewhere in the UK, Scotland's climate is influenced by the surrounding waters, which have a moderating effect that tends to result in relatively mild winters and warm summers. However, Scotland tends to be colder and cloudier than the rest of the UK, and in parts, such as the Highlands, winters can be far more severe, with heavy snowfalls, whilst the west coast experiences more favourable conditions on account of the effects of the Gulf Stream.

Norway
Considering its northerly position, Norway has a temperate climate, with mild winters and cool summers, particularly along the coast, which benefits from the warming effects of the North Atlantic Drift; a current that extends from the Gulf Stream. Summers are colder and winters more severe inland, with increased snowfall at higher elevations, although the west coast experiences greater rainfall.

The average annual precipitation is 990 millimetres/39 inches, whilst the average winter temperature is around 2°C/27°F, and the average summer temperature ranges between around 13–18°C/57–65°F.

Sweden
Sweden's climate is generally continental and temperate although, like neighbouring Finland and Norway, it is subject to maritime influences and may be considered relatively moderate, particularly in the south. Winters may be long and quite severe in the sub-arctic north, however, and the summers tend to be short and mild throughout the country.

In the north, average temperatures range from about −11 to 15°C/12 to 59°F from February, the coldest month, to July, whilst in the south the average

Western Europe

Austria

Being largely mountainous, the Austrian climate varies according to regional topography, generally becoming colder at higher elevations. However in winter, cold air may descend into the valleys. The country is subject to three main influences; the Atlantic, Mediterranean and the continental, and also to warm local winds, such as the föhn, which typically help to bring about the spring thaws. Generally, winters are cold, summers warm and wet, and spring and autumn are mild.

Annual temperatures average between about 7–9°C/44–48°F overall, but in the warmest month, July, may average between 18–20°C/64–68°F in major cities such as Vienna. Rainfall also tends to be highest in summer, but averages about 13 millimetres/½ inch per month.

Belgium

The climate is generally temperate, with maritime influences in coastal areas giving rise to mild, humid, often foggy conditions. In the interior, summers may be hot, whilst winters are typically cold, with frost and heavy rains experienced in upland areas, although precipitation is fairly constant throughout the year.

The average annual rainfall throughout Belgium is around 699 millimetres/28 inches, and the average temperature about 8°C/47°F, whilst in Brussels the average temperature ranges between -1 to 4°C/30 to 39°F in January, and 12–23°C/54–73°F in July.

Average temperatures range from about 2°C/35°F in January, to around 16°C/60°F in July, and the average annual precipitation ranges from over 3000 millimetres/120 inches in the western Highlands, to less than 800 millimetres/32 inches in parts of the east.

Wales

The climate of Wales is temperate and moderate, with warm, often wet summers, and mild winters, which result from a combination of the hilly terrain, its proximity to the Atlantic Ocean, and the prevailing westerly winds that bring moisture inland. Precipitation varies regionally however, increasing with altitude, so that most is experienced in the mountainous Snowdonia area, where heavy winter snowfalls are also common.

In the east and in coastal areas, annual precipitation may average less than 1000 millimetres/40 inches, but in Snowdonia it can exceed 3000 millimetres/120 inches. The average January temperature for Wales is around 5°C/41°F, rising to about 16°C/60°F in July.

Above: Rainbow over Kyeakin, Isle of Skye, Scotland. Average annual rainfall in the west of Scotland is approximately 3000 millimetres/120 inches.

Right: The temperate climate of England is ideal for deciduous trees, such as the beech. Bluebells carpet the woodlands in early spring when the new leaves are not yet thick enough to cut out all the sunlight.

Opposite: Alesund, in Norway enjoys a relatively mild climate thanks to the warming effects of the North Atlantic Drift, an offshoot of the Gulf Stream.

France

The Climate of France may generally be described as temperate, but there are distinct regional variations, from the Mediterranean south, where summers are hot and dry, and winters are warm and humid, to the north and west coasts, where prevailing winds and Atlantic currents have a moderating effect, and the interior and eastern parts, where the climate is more continental and winters may be cold, with long periods of snow cover, and the summers are warm or hot. The southeast is also subject to the Mistral, a cold north wind that descends from the Alps and the plateau of the Massif Central.

   The average temperature ranges from about 5°C/41°F in January to 22°C/72°F in July, and the average annual precipitation varies from around 620 millimetres/24 inches in Paris, to 550 millimetres/22 inches in Marseille. However in mountainous regions that experience heavy snowfalls, the average may be far greater.

Germany

Germany's climate is temperate, with maritime influences, particularly in the northwest, where winters may be mild, but further inland there may be more extreme seasonal variation, with colder winters and warmer summers. Eastern parts also tend to experience fairly moderate conditions, whilst at higher elevations in the interior and in the south, an alpine climate is experienced, with increased precipitation, colder temperatures, and often prolonged snow-cover.

   According to region, the average annual temperature ranges from about -6 to 1°C/21 to 34°F in January, and from about 16–20°C/61–68°F in July, whilst precipitation varies from about 710 millimetres/ 28 inches in the northern lowlands, to 1900 millimetres/ 78 inches in the south, where much of it falls as snow.

Liechtenstein

Liechtenstein is landlocked, and lies partly in the flood plain of the upper Rhine Valley, and partly in the Alps, giving it both a continental and an alpine climate. Winters can be cold and cloudy with snow and rain, but may often also be relatively mild, moderated by warm, south föhn winds, which hasten the arrival of spring. Summers may be cool or warm, and are often humid.

   Average temperatures in the capital Vaduz, range from about 1°C/34°F in January, to about 18°C/64°F in July, whilst precipitation may range from just 150 millimetres/6 inches in some parts to about 1200 millimetres/47 inches at higher elevations in the alpine regions.

Left: Iced branches are seen in a snow-covered valley near the southern German town of Koenigssee. Germany's climate is moderate and is generally without sustained periods of cold or heat.

Opposite below: Austrian summers are warm and sunny, with pleasant mountain breezes – perfect walking conditions.

Luxembourg

The climate of Luxembourg is generally temperate and moderate, and transitional between that of continental eastern Europe and the more maritime climate of northwest Europe. Summers tend to be relatively cool and winters mild, although in the north of the country, particularly at higher elevations in the Ardennes, the winters are typically more severe.

Precipitation is greatest in the southwest, but overall the annual average is around 762 millimetres/ 30 inches. Temperatures may fall below 0°C/32°F in winter and reach above 30°C/86°F in summer, with a national annual average of about 10°C/50°F.

Monaco

Monaco's climate is Mediterranean, with hot summers, mild winters, and around 300 days of sunshine annually, uninterrupted by rain. The hottest months are July and August, and the coolest are January and February.

Average annual precipitation is around 730 millimetres/29 inches, which tends to fall mainly in October and November. Average temperatures range from about 8–12°C/46–54°F in January, to between about 22–26°C/72–79°F in August.

The Netherlands

The climate of the Netherlands is temperate and continental, but is largely moderated by the maritime influences of the Atlantic Ocean and the North Sea,
varying only slightly inland. It is frequently calm and cloudy, with relatively even rainfall throughout the year. In winter, cold winds and moist air can reduce temperatures, particularly in coastal areas, as can freezing Arctic and Siberian air masses, whilst conditions are often mild from about May to September.

In January the average temperature ranges from about –1 to 4°C/30 to 39°F, and in July, from around 13–22°C/55–72°F. The annual precipitation averages about 760 millimetres/30 inches.

Switzerland

Switzerland's climate is temperate, but conditions vary regionally according to the country's topography, which ranges from low-lying valleys and plateaus to the mountains of the Alps, where there are large glaciers and permanent snow-cover, and temperature decreases rapidly with altitude. Winter temperatures are typically freezing throughout the country, however, whilst summers may be warm, particularly at low elevations. In the summer, bise winds may move through the valleys, warming by day and cooling by night, and in the winter, warm, dry föhn winds may move over the Alps.

In January, the temperature averages around –2°C/28°F, whilst in July it averages about 16°C/61°F, and the average annual precipitation is about 986 millimetres/39 inches in the north of the country and 914 millimetres/36 inches in the south.

Eastern Europe

Belarus

The climate of Belarus is temperate and continental, moderated by the both the Atlantic Ocean and Baltic Sea. It is generally cool and humid, with warm, moist summers, cold winters, and fairly high precipitation, particularly in the interior.

Average annual temperatures range from around 18°C/64°F in July, to –7°C/20°F in January, although much colder temperatures are sometimes experienced in the north of the country. Annual precipitation ranges between about 546 millimetres/ 22 inches to 700 millimetres/30 inches.

Bulgaria

The climate is generally continental, with hot summers and cold winters, although in the south, conditions are more Mediterranean, experiencing drier summers and mild, moist winters. Winters also tend to be warmer along the Black Sea coast, but Russian northeast winds frequently bring colder periods.

Average rainfall is around 630 millimetres/ 25 inches per year, with February being the driest month and May the wettest throughout much of the country. However in the south, more rainfall may be experienced in autumn and winter.

The average temperature in January is 2°C/36°F, and in July 21°C/70°F.

Czech Republic

The climate is generally temperate and continental, with warm, humid summers and cold, fairly dry winters, particularly in the flat interior regions, although in the mountainous areas of the north, the winters are more severe, often bringing heavy rains, and persistent snowfalls, whilst in the south, summers may be hot, and the winters mild.

The average temperature range for January is around –4 to 1°C/25 to 34°F, and for July, about 14 to 23°C/57 to 73°F, with average annual precipitation of around 693 millimetres/27 inches.

Hungary

Hungary has a temperate, continental climate, with marked seasonal variations. The spring is often warm and mild, the summer is generally hot and wet with heavy thunderstorms, autumn is typically dry, and winter may be cold and cloudy with frequent snow showers.

Average temperatures range between a low in January of around –1°C/30°F, to a high in July of about 21°C/70°F. Annual precipitation averages around 640 millimetres/25 inches.

Moldova

The climate is generally temperate and continental, but influenced to an extent by the Black Sea. Winters tend to be moderate, mild and dry, whilst summers are warm and moist. Precipitation is variable, and is usually highest in the north of the country at higher altitudes, whilst droughts are a frequent occurrence in the south.

Annual rainfall ranges from about 610 millimetres/24 inches in the north, to around 350 millimetres/14 inches in the south, whilst average temperatures range from about –5°C/23°F in January, to around 24°C/75°F in July.

Poland

Poland's climate is generally continental, with relatively mild, wet summers, during which thunderstorms are common, and quite severe winters, when snow may cover the ground for a month or more, particularly in the north and east. The Baltic Sea however, has a moderating influence in western parts.

Most rainfall occurs in summer, although there may be quite a contrast between the amount received in lowland and mountainous regions, and the overall annual precipitation averages around

610 millimetres/24 inches. Average temperatures range from between about –5 to 0°C/23 to 32°F in January, and from 17 to 24°C/63–75°F in July.

Romania

The climate of Romania is temperate and continental, with long, cold and snowy winters, and warm or even hot summers. In the mountainous regions, winter temperatures are lower, and a severe northeast wind, the crivat, blows in from Russia, whilst in summer the Carpathian Mountains may be cool and moist, although the prevailing winds affecting lowland regions are typically warm and dry and drought conditions are often experienced in these areas.

Temperatures range from about –7°C/19°F in January to around 30°C/86°F in June, and annual precipitation averages about 637 millimetres/25 inches.

Slovakia

Slovakia has a temperate, frequently humid, continental climate, with cold winters and warm summers. The eastern lowlands tend to experience hot summers, with less precipitation, whilst winters can be harsh in the High Tatras and Carpathian Mountains, where precipitation is also heaviest.

Annual rainfall can be as high as 1000 to 2000 millimetres/40 to 80 inches in mountainous regions, but averages around 490 millimetres/ 19 inches for the country overall. The average temperature is about –1°C/30°F in January and around 21°C/70°F in July.

Ukraine

Ukraine is essentially temperate and continental, with four distinct seasons, long, cold winters and hot summers, although on the southern Crimean coast it experiences Mediterranean conditions, with milder winters and drier summers. In the east of the country summers may also be warmed and winters cooled by air moving from the steppes of Asia.

In the north, average temperatures range from –8 to 29°C/18 to 85°F from January to June, and from –2 to 32°C/28 to 90°F in the south, and the overall average annual precipitation is around 500 millimetres/20 inches. However, precipitation in mountainous areas and the north may often be double that of the south.

Southern Europe

Albania

The climate is generally temperate, although low-lying coastal areas benefit from Mediterranean conditions, with hot dry summers and mild winters. Most precipitation occurs in winter, when there may be thunderstorms in the lowlands and heavy snowfalls in the mountainous regions, although the mountains tend to receive more rain in summer.

The hottest month is August, with temperatures averaging between 17–31°C/63–88°F, and the driest month is July, with an average rainfall of about 32 millimetres/1¼ inches. January is the coldest

Opposite above: Ice fishing in Belarus – several men catch and ice fish on a frozen lake. Belarus is a landlocked country which experiences cold winters and warm and moist summers.

Right: In the Carpathian mountains, which lie between Poland and Slovakia, the annual rainfall is very heavy. In winter the slopes are thick with snow.

month, with an average temperature between 2 and 12°C/36–54°F, whilst December is the wettest, with about 210 millimetres/8½ inches of rain.

Andorra

Being located on the southern slopes of the Pyrenees mountain range, Andorra's climate may be described as alpine, with mild, or warm, dry summers, depending on elevation, and typically severe winters, as a result of which valleys in the north of the country may be snowbound for over half of the year.

Average temperatures range between –1 to 6°C/30 to 43°F in January, to 12–26°C/54–79°F in July, with an average annual precipitation of about 808 millimetre/32 inches.

Bosnia and Herzegovina

The climate of Bosnia and Herzegovina varies regionally, from a modified, or sub-Mediterranean climate in the south of the country, where summers are warm and sunny, and winters are relatively mild, to a more continental climate in the north and interior, becoming more alpine at higher elevations, with longer, colder winters, which may be quite severe.

The average temperature in January is around 0°C/32°F, and in the warmest month, July, about 22°C/72°F. The average annual rainfall is around 625 millimetre/25 inches.

Croatia

Croatia experiences distinct regional differences in climate. A Mediterranean climate of hot, dry summers and mild winters is experienced along the coast south of Split, a continental climate, with hot summers and cold winters is characteristic of the Pannonian Plain, mild summers and winters with high precipitation are found in the Dinaric mountains, and an Adriatic climate is experienced along much of the coast, with long warm summers and short, dry winters, influenced by the Bura winds.

Annual rainfall averages about 870 millimetre/34 inches overall, whilst temperatures range from about 0°C/32°F in inland areas in January, to 24°C/75°F in July, and along the coast from 9–25°C/48–77°F.

Greece

Greece experiences a Mediterranean climate, with a great deal of sunshine throughout the year, long, hot, dry summers, and mild, wet winters. However, regional variations occur, with a typically drier climate in the south and east than in the north and west.

Average annual precipitation ranges from about 500–1210 millimetres/20–48 inches in the north, and from around 380–810millimetres/15–32 inches in the south. Summer temperatures average between 29 and 35°C/84–95°F in July and August,

Left: Low lying fog over fields in Italy, where the climate is quite diverse with five main zones. The Alpine zone has cold winters and warm summers, and is quite wet, the Po Plain has a continental climate with quite cold winters and warm, mild summers, the western Tyrrhenian area has cool winters and warm summers, the Adriatic coast has warm summers but cold winters, and the southern, Mediterranean has mild, wet winters and hot, dry summers.

Opposite above: Greece enjoys hot summers and mild winters with many sunny days, but the south and east can be very dry.

but in coastal areas, a cooling north wind known as the Etesian, frequently blows from the Aegean Sea. In the coldest months of January and February, average temperatures may range from below freezing in the north, to between 5 and 10°C/41–50°F at the coast, and from 0 to 5°C/32–41°F in the interior, but sunny 'Halcyon days' continue to be experienced.

Italy

Italy's climate is temperate but diverse, varying from region to region, and with five main zones: the Alpine zone, with cold winters, warm summers, and high precipitation; the Po Plain, which experiences a continental climate with quite cold winters and warm, mild summers; the western Tyrrhenian area, where winters are cool and summers warm; the Adriatic coast, with warm summers but cold winters, when temperatures are reduced by the bora, a cold north-easterly wind; and the southern, Mediterranean area, which experiences mild, wet winters and hot, dry summers, subject to hot sirocco winds from Africa.

The average annual temperature ranges between 11 and 19°C/52–66°F, whilst precipitation ranges from about 1525 millimetres/60 inches in parts of the north, to about 500 millimetres/20 inches in parts of the south.

Macedonia

The climate of Macedonia is essentially continental, with hot, dry summers, particularly in the mountainous regions, dry autumns, and cold winters. At higher elevations there may be heavy winter snowfalls, whilst low-lying regions tend to be milder throughout the year.

The average annual precipitation is around 510 millimetres/20 inches, and average temperatures for January and July respectively, range from approximately –20 to 0°C/–4 to 32°F, and from 20 to 23°C/70 to 73°F.

Malta

Malta's climate is Mediterranean, with relatively hot, dry, sunny summers and mild, wet winters, which may be accompanied by periods of foggy weather.

Most rainfall tends to occur between the months of October and January, and the average

annual precipitation is around 560 millimetres/ 21 inches, with December being the wettest month of the year and July the driest. Average temperatures range between about 9°C/48°F in winter, to 31°C/88°F in summer, with January being the coldest month and August the hottest.

Montenegro

Montenegro's climate is generally temperate, with somewhat continental conditions being experienced

Above: Vineyards in Solvenia. Away from the coast the moderate climate is quite continental and is ideal for growing grapes.

Opposite below: The extreme south of Spain experiences the highest winter temperatures of the European mainland, with an average of around 14°C/57°F.

in the interior, although a Mediterranean climate prevails along the Adriatic coast, with warm, dry summers and mild winters. Temperatures along the coast range from about 4–12°C/39–54°F in January, and from about 19–29°C/66–84°F in July, but at higher elevations in the mountains, temperature ranges for January and July are between about –7 and 2°C/19 and 36°F and 9–23°C/48–73°F respectively. In terms of precipitation, the coastal mountains experience one of the highest average annual rainfalls in Europe.

Portugal

Portugal's climate varies quite considerably according to region and topography, from a temperate maritime climate in the north, with mild, wet winters and cool summers, to mild, wet winters and hot summers in central parts. In the south of the country meanwhile, the climate is much drier, and tends to be hotter still.

Overall, the average annual temperature is around 16°C/61°F. Rainfall tends to be heaviest in the north, where there may also be snowfalls in winter, with precipitation at 1250–1500 millimetres/ 50–60 inches per year, whilst in the south it is around just 500 millimetres/20 inches.

San Marino

San Marino shares its climate with that of surrounding northeast Italy, being Mediterranean, with warm, sunny summers and cool or mild winters, although temperatures may fall below freezing during the winter months.

In summer, average temperatures range between about 20–30°C/68–86°F, whilst in winter, temperatures generally fall no lower than –6°C/21°F. Precipitation is moderate, with between about 560 and 800 millimetres/22 and 32 inches being received throughout the year.

Serbia

Serbia has a moderate continental climate, with cold winters and fairly hot, humid summers, particularly in the north, whilst further south, the country experiences more Adriatic or Mediterranean conditions, with drier summers. In the interior there may be heavy snowfalls in winter, especially at higher elevations. Temperatures range from an average of about 2–3°C/36–37°F in January, up to about 23°C/73°F in July.

Rainfall is well distributed throughout the year in the north of the country, whilst in the south more tends to be received during the summer, often during thunderstorms. Average annual precipitation ranges from about 1000–1500 millimetres/20–39 inches in lower-lying areas, up to around 3800 millimetres/152 inches in the mountainous regions.

Slovenia

In coastal areas, Slovenia experiences a Mediterranean climate, but inland the climate is more continental, with warm or hot summers, and cold winters, particularly in the eastern valleys and plateaus. In the mountainous regions of the north, the summers may be quite wet, and the winters more severe.

Average temperatures range from about 0–2°C/32–36°F in January, the coldest month, and from about 18–19/64–66°F in July, the warmest. Average annual precipitation varies from about 800 millimetres/32 inches in drier, low-lying areas to around 2000 millimetres/79 inches in the mountains.

Spain

Spain's climate varies quite considerably, from the relatively cool, humid northern coast, to the continental central plateau, where summers are hot and winters are cold, to the Mediterranean south and east, and the sub-tropical of the extreme south, which experiences the highest winter temperatures of the European mainland, with an average of around 14°C/57°F.

Elsewhere, temperatures range from about 10–19°C/50–66°F between winter and summer in Bilbao, and around 8–23°C/46–73°F in Madrid, whilst the average annual rainfall is around 890 millimetres/35 inches or greater in the north, and between about 310 millimetres–790 millimetres/ 16–31 inches in parts of the south.

AFRICA

Western Africa

On the west coast, from Senegal to Sierra Leone, conditions are tropical, with a hot and humid rainy season from about June to November, followed by a cooler, dry season, which lasts for the remainder of the year. During the rainy season, temperatures average between about 26–32°C/79–90°F, but during the dry season, they may range from as low as 18°C/65°F to about 27°C/81°F. However, temperatures are typically slightly cooler and precipitation is often higher along the coast than in the northern and interior parts of these countries. In Senegal and The Gambia for example, harmattan winds from the Sahara may raise temperatures up to about 38°C/100°F at times. Annual rainfall ranges from about 1400–2000 millimetres/55–80 inches, although in the highlands of Guinea it may exceed 2800 millimetres/110 inches and on the coast of Sierra Leone, be as high as 3800 millimetres/150 inches.

Similar conditions are experienced across the northern parts of Liberia, Ivory Coast, Ghana, Togo and Benin, but further south the climate is equatorial, with a long rainy season from about March to July, followed by a short dry season until about mid-September, a short wet season until mid-November, and then a long dry season until March. In Nigeria, conditions are highly variable, with persistent rains throughout the year in southern coastal areas bringing up to 4000 millimetres/160 inches of rain, whilst in the middle of the country the rainy season lasts from about April to October, and delivers about 1000 to 1500 millimetres/40 to 60 inches. In the far north meanwhile, the rainy season is far shorter, and rainfall often averages below 500 millimetres/20 inches annually. Temperatures can average between about 25–28°C/77–82°F throughout the year, but in places may dip to below 12°C/54°F at night, and exceed 38°C/100°F during the day.

In Mali and Niger, in the West African interior, the climate is hot and dry, with desert conditions in the north, with extremely high temperatures and precipitation of less than 127 millimetres/5 inches per year, through the semi-arid Sahel, to southern parts, which experience a short wet season from about June to September, which may bring between 813–1400 millimetres/32–55 inches of rain.

Middle Africa

As in much of the continent, conditions in Middle Africa range from the almost continuously hot and arid, as in the deserts of northern Chad for example, through regions with three distinct seasons, such as in the middle of that same country, which experiences

Left: The rippled floor of the Namib desert, where the average annual precipitation may be just 50 millimetres/2 inches, in part due to the cold Benguela Current.

Opposite above: Kenya enjoys a tropical climate. It is hot and humid at the coast, temperate inland and very dry in the north and northeast parts of the country.

Opposite below: Altitude affects weather conditions, so despite its southern position some mountainous areas of Africa have cold winters and it may even snow.

a hot, dry season from about March to July, a rainy season until October, and then a cool season for the rest of the year, into tropical hot and humid countries, such as the Central African Republic and Cameroon, where there may be a rainy season and an extended dry season, or even two dry seasons. Further south into equatorial regions, such as Equatorial Guinea and parts of the Congo, there may be persistent rains throughout much of the year at the equator, particularly on some mountain slopes, or a distinct rainy and dry season, or even two wet and two dry seasons in regions both north and south of the equatorial divide. Additionally, at high elevations, such as in the mountains of the Democratic Republic of the Congo, there may be heavy snows, and at lower altitudes in the mountains and foothills the climate may be alpine and essentially temperate. Further south still, moving into Angola, there is considerable variation in climate, both along the coast from the north to the south, and between coastal areas and the interior. In the north, the climate is somewhat more moist and humid, receiving as much as 1500 millimetres/59 inches of rain, becoming increasingly arid towards the south and the Namib Desert, where annual precipitation may be just 50 millimetres/2 inches – in part due to the cold Benguela Current – although the central plateau actually experiences lower temperatures. Overall, the average temperature for Angola is around 20°C/68°F.

Elsewhere, across the region of Middle Africa as a whole, average temperatures tend to range seasonally from about 14–33°C/61–91°F, although temperatures may reach as low as −12°C/10°F in parts of Chad during winter nights in the desert, to

highs of 50°C/122°F, both there and in the far north of Cameroon. Differences in annual rainfall can be similarly extreme, ranging from an average of 25 millimetres/1 inch in northern Chad, to 10,920 millimetres/430 inches.

Eastern Africa

The conditions in Eastern Africa range from incredibly hot and Arid, particularly in the north, in countries such as in Eritrea, Djibouti and parts of Somalia, through several countries, including Ethiopia, Uganda and Rwanda, where altitude has a considerably moderating effect, to various tropical and subtropical areas that may be prone to both seasonal flooding and droughts.

Some of the highest recorded temperatures come from the Danakil Depression in southern Eritrea, where highs of 50°C/122°F are not uncommon, and little rainfall is received. Further inland, in the highlands, it is cooler and wetter however. Djibouti and Somalia also record high temperatures, and often experience droughts, but whilst the average annual rainfall is typically less than 130 millimetres/ 5 inches in Djibouti, rains can be unpredictable and flash floods may also occur.

From Ethiopia south to Tanzania conditions vary, but most countries are subjected to both a long and a short rainy season. However, throughout this region, topography and altitude have some of the most notable effects on climate. For example, despite being equatorial, countries such as Uganda experience relatively mild conditions, with an average annual temperature range of between about 16–29°C/60–84°F.

In Zambia, Zimbabwe, Malawi, and Mozambique, there is a rainy season from around November to March or April, followed by a cool dry season from May to August, and in most places, a hot dry season from August until the major rains begin again in November. However, in Mozambique, the rainy season is followed by a prolonged dry season from April until about October, during which drought can be a considerable problem, and the country also often experiences severe flooding during the rainy season. The average annual rainfall in Mozambique ranges from about just 300 millimetres/12 inches in parts of the south, to over 1420 millimetres/56 inches in more central parts, whilst in parts of Malawi and Madagascar, annual precipitation may be as great as 1780 millimetres/70 inches and 3050 millimetres/120 inches respectively, and Madagascar is occasionally subjected to devastating cyclones. Temperatures in the region range from as low as 5°C/41°F during the winter nights in Zambia, to as high as 38°C/100°F during summer days in Zimbabwe.

Southern Africa

The climate of southern Africa varies from extremely hot and arid, in the Namibian deserts, through semi-arid, subtropical conditions in neighbouring Botswana, to the variable, but generally milder, temperate or near-temperate conditions, which are found into the mid-latitudes across parts of South Africa, Lesotho and Swaziland.

With its deserts that extend right up to the coast, Namibia has the driest climate on the African continent, and although it may receive an average annual rainfall of between 560–700 millimetres/22–28 inches in northernmost parts, this figure decreases to about 350 millimetres/14 inches in some central areas, 150 millimetres/6 inches in the south, and around 50 millimetres/2 inches at the coast. Precipitation tends to occur between November and March, but in many areas the rains frequently fail to occur at all, and some places have been without rain for decades. At the coast, much of the moisture received is in the form of blanketing fog which may bring slight drizzle. Temperatures in Namibia range from about 23–29°C/73–85°F during January in summer, to an average winter temperature of about 13°C/55°F in July.

In Botswana, the rainy season may be slightly longer, with most precipitation occurring between December and April, and averaging about 450 millimetres/18 inches per year, but droughts are common here too. In winter, temperatures may drop to below 10°C/50°F, particularly at night, but in summer, they range from about 22–33°C/72–91°F, with October usually being the hottest month.

South Africa is generally warm and temperate, with its climate being moderated by maritime influences and altitude, but there are marked variations in temperature between the east and west coasts. In the east, the warm Agulhas Current ensures that temperatures range from about 11–22°C/52–72°F in winter, and from around 21–27°C/69–81°F in summer, whilst in the west, the cold Benguela Current brings average winter temperatures of about 7°C/45°F, and average summer temperatures of around just 12°C/54°F.

Rainfall is typically varied and unpredictable across much of South Africa, with periods of both drought and flooding, and whilst the average annual precipitation for the country is approximately 460 millimetres/18 inches, the west is generally much more arid, with a typical average of below 200 millimetres/8 inches.

ASIA

Russia

Russia is a vast country that stretches from northeastern Europe across northern Asia, and most of the landmass experiences a continental climate, with short, warm, or hot summers, and long, very cold winters, which are often severe on account of the lack of maritime influences and the generally high latitude. In the far north, which experiences an Arctic climate, the Arctic Ocean freezes along much of the coast, whilst in the south, mountain ranges tend to present a barrier to much of the warm, tropical air that might otherwise extend into the country from southern Asia. Similarly, in the east, warm air from the Pacific rarely reaches far inland, on account of the prevailing westerly winds, and in winter, the cooling high pressure system over Mongolia further reduces temperatures. The greatest maritime influences spread from the Atlantic in the west, bringing warm, moist air in summer, when most rainfall tends to occur over much of the country, but overall, as the air is so cool, precipitation is relatively low throughout the year, despite frequently overcast skies, particularly in the north.

The average annual precipitation ranges from between about 640–760 millimetres/25–30 inches in the European, western regions, to between 508 and 813 millimetres/20 and 32 inches across Siberia and far eastern Russia, but may be as high as 1016 to 2000 millimetres/40 to 79 inches at higher elevations in mountainous regions, and as low as 50 millimetres/2 inches in central Asian parts.

Temperatures range from an average of 13°C/57°F at Sochi, situated on the Black Sea coast, and 4°C/40°F in Moscow, to –12°C/9°F at Tiksi in northern Siberia. However, the highest temperature ever recorded in the country is 38°C/100°F, while temperatures as low as –70°C/–90°F have been recorded at Verkhoyansk and Oymyakon; the coldest of any inhabited location.

Central Asia

In general, the climate of Central Asia is highly continental. It ranges from the semi-arid, in the steppe grasslands that are a major feature throughout much of the region, to arid, in the deserts that

Above: A Siberian tiger in deep snow. In Siberia the temperature averages –12°C/9°F.

Opposite: On the coast of Namibia, much of the moisture received is in the form of a blanketing fog, which may bring slight drizzle.

Below: Yurts in the snow in Mongolia.

dominate Turkmenistan and Uzbekistan. Some areas are classed as extremely cold alpine – which is essentially equivalent to polar – such as the Tien Shan mountains of Kyrgyzstan, and the Pamir and Alai mountain ranges in Tajikistan, where glaciers and permanent snow-cover are major features.

Overall, there is very little precipitation in Central Asia, although there may be great variation in rainfall depending on the topography, with less than 100 millimetres/4 inches typically being received annually in desert areas, to an average of over 1600 millimetres/64 inches in some of the more mountainous parts.

Similarly there may be extreme variations in temperature, not only according to location and elevation, but also seasonally, and even between day and night, particularly in the deserts. Average temperatures range from about –5°C/23°F in January to between about 26–32°C/79–90°F in July, although in places, winter temperatures can fall as low as –45°C/–49°F, whilst in others, such as in the Kara Kum desert in Turkmenistan, daytime summer temperatures can easily reach 50°C/122°F.

Southwest Asia and the Middle East
Although the climate of Southwest Asia and the Middle East is generally characterized as semi-arid or arid, with long, hot summers and mild, often wet winters, large regional variations occur. These also encompass temperate, continental, Mediterranean, alpine, and subtropical climates, and a wide range

of temperatures and degrees of humidity and precipitation, which may vary according to such factors as topography and maritime influences. Even within countries there can be marked seasonal and geographical weather differences or climate types, and extremely hot days may be followed by warm or even cold nights. Coastal regions often experience milder, rainy conditions in winter, whilst mountainous areas often experience much colder temperatures and increased precipitation, including snow. Inland areas at lower elevations, meanwhile, may experience continental climates, with hot summers and fairly cold or cool winters, whilst there are desert regions,

Above: The monsoon season can bring precipitation of up to 7800 millimetres/312 inches in parts of Bhutan.

Left: Nearly one-fifth of the surface of the globe can be described as desert. Sand dunes such as these are characteristic of Saharan Africa and Saudi Arabia. However, other desert areas can support specialized vegetation, particularly after rains.

Opposite above: Across much of Asia the wet or rainy season results from the southwest monsoon, which brings heavy rain from May or June to November or December, following which, cooler conditions often dominate until perhaps March or April.

on the Arabian Peninsula for example, which may receive almost no precipitation, and where temperatures may remain relatively high throughout the year. However, there are also parts of the Arabian Peninsula, such as the southern and southwestern coasts, which are affected by monsoons, and in fact, much of Southwest Asia and the Middle East can be described as having a hot, dry season, and a cool, rainy season.

Cyprus, Turkey, Syria, Lebanon, Palestine and Israel, all benefit from a largely Mediterranean climate, with average winter and summer temperatures of between about 5–23°C/41–73°F respectively, whilst the climates of Armenia and Iraq, for example, are more continental, and those of Bahrain, Kuwait, Saudi Arabia, the United Arab Emirates and Yemen are far more arid, with average summer temperatures frequently in excess of 30°C/86°F, and winter temperatures as high as 19°C/66°F in places.

The highest precipitation for the region as a whole occurs in the coastal regions of the Mediterranean and Caspian Seas, and the mountainous regions of Turkey, Iraq and Lebanon.

Southern Asia

From Iran in the west, across Afghanistan, Pakistan, and Nepal, to Bhutan and Bangladesh in the east, and south through India to Sri Lanka and the Maldives, a wide range of climatic conditions may be encountered, from the extremely hot and arid, to freezing alpine, and from the temperate to subtropical and tropical monsoon. It should also be noted that there may be great variation not only from country to country, but within countries themselves. For example, although Iran may be said to have a generally continental climate, with cold winters and hot summers, at higher elevations winters may be more severe and summers milder, with greater precipitation, whilst the interior is arid, and some coastal areas almost constantly humid, with a subtropical climate.

Similarly, in Afghanistan conditions are highly variable, with average winter temperatures of around –9°C/15°F in places, and summer highs of perhaps 49°C/120°F. Even in the course of a day, temperatures may range from below freezing during the night and at dawn, to around 40°C/104°F at midday. Most precipitation in Afghanistan falls as snow in winter and spring.

In Pakistan, Nepal and Bhutan, winters can also be severe in mountainous areas, with permanent snow-cover at some elevations, although more southerly parts tend to experience subtropical conditions, with a hot dry season, when temperatures may reach up to 49°C/120°F in parts of Pakistan, before the monsoon rains break, bringing precipitation of up to 7800 millimetres/312 inches in parts of Bhutan.

Aside from at height in the Himalayas, the monsoon also dominates the Indian climate, as well as those of Bangladesh and Sri Lanka, which are essentially tropical, and experience wet and dry seasons annually. The wet or rainy season results from the southwest monsoon, which brings heavy rain to much of the subcontinent throughout the period from May or June to November or December, following which, cooler conditions often dominate until perhaps March or April as a result of the northeast monsoon, followed by hot dry weather until the rains return. The weather in India can be extreme – the village of Mawsynram in northeastern India for example, experiences the highest average annual rainfall on Earth, at about 11,873 millimetres/467 inches, whilst there are villages in parts of Rajasthan that receive less than 130 millimetres/5 inches. Temperatures meanwhile may range from well below freezing in mountainous areas in winter, to exceed 52°C/125°F in the interior at the height of summer.

Eastern Asia

Climatic conditions in Eastern Asia vary considerably, from the arid continental climate of Mongolia, which experiences extremely cold winters, through the largely temperate, but highly varied climates of China, to the mainly continental regions of North and South Korea, and the sub-tropical and tropical areas of Hong Kong, Japan and Taiwan. These latter areas experience monsoon rains and frequent tropical cyclones, or typhoons in summer and autumn.

The 'great Siberian high', the pressure system that can lower winter temperatures to $-30°C/-22°F$ in Mongolia, also affects the surrounding regions, causing cold, dry winds to blow across northern China, North and South Korea, and Northern Japan in winter, frequently reducing temperatures to well below freezing. Temperatures may be further reduced by the Okhotsk or Oyashio Current, which flows down from the north into the Sea of Japan.

Conversely in summer, warm air from the southern monsoon regions and precipitation from the remnants of tropical cyclones may reach Mongolia, bringing most of the annual rainfall to areas such as the Gobi desert, which typically experiences just 100 millimetres/4 inches per year. The monsoons also bring heavy winds, high humidity and most of the rains to southern China, Hong Kong, Japan and Taiwan in summer, and there are generally several typhoons each year in these areas, which may also reach South Korea in late summer.

South East Asia

The climate of South East Asia is perhaps best described as tropical. It is influenced by maritime winds that emanate from both the Indian Ocean and the South China Sea, and is characterized by high temperatures, high humidity and copious precipitation in the form of monsoon rains. Depending on the exact location, there may be one or two rainy seasons each year: a summer monsoon, caused by southwest winds and a winter monsoon, brought by the northeast trade winds. These wet seasons are usually interspersed with a cool, dry season, a hot dry season, or perhaps both. However, some South East Asian countries, such as Brunei and Singapore for example, lack a dry season, and although they may receive increased precipitation during the monsoon, annual rainfall, humidity and temperature may be consistently high throughout the year.

In Myanmar (formerly Burma), Thailand, Laos, Cambodia, Vietnam and the Philippines, the wet season usually lasts from around April, May or June, to about October or November, during which over 2500 millimetre/100 inches of precipitation may be received in some areas, and typhoons may threaten the Philippines and Vietnam particularly. In Indonesia, Malaysia and Timor-Leste/East Timor, meanwhile, the heaviest rains tend to occur between November or December, and March or April. In many parts of South East Asia, the average humidity stands at about 80 per cent, whilst annual precipitation may frequently exceed 3000 millimetres/150 inches – although in mountainous regions in Cambodia, Malaysia and Myanmar it may be as high as 5080 millimetre/ 200 inches. Myanmar also boasts the tallest, and only snow-capped peaks in South East Asia.

Temperatures in the region probably average between about $24–30°C/75–86°F$ throughout the year, but may range from as low as $5°C/41°F$, during the dry season in Hanoi, northern Vietnam for example, to up to $38–40°C/97–104°F$, during the hot, dry season in Laos.

Above: An image of Japan in the spring: cherry trees heavy with blossom and the snow-covered Mount Iwaki in the background.

Opposite above: Inland in the Northern Territories of Australia the climate is semi-arid.

AUSTRALASIA AND PACIFIC ISLAND GROUPS

Australia

Generally, Australia's climate is warm, dry and sunny, but the continent experiences a wide range of conditions, from temperate in the south, with relatively moderate temperatures, to hot, semi-arid and arid in the interior, humid in the east, and subtropical and tropical, with monsoon rains, in parts of the north.

Winters are typically mild or warm, with average July temperatures ranging from about 9°C/48°F in Melbourne, which is located in the southeast of the country, and from 8–16°C/46– 61°F in Sydney, to 25°C/77°F in the north at Darwin. Temperatures may often dip below freezing at night on account of the usually clear skies, but frost and snow are relatively rare except at higher elevations.

Average temperatures for January, during summer, range from 20°C/68°F in Melbourne, and 18–26°C/64–79°F in Sydney, to 30°C/86°F in Darwin, although average summer highs of between 33–38°C/91–100°F are common throughout much of the country. In the arid outback of the interior, temperatures may frequently rise above 46°C/115°F.

Rainfall also varies considerably from region to region, but overall, it is somewhat sparse, averaging about 420 millimetres/17 inches annually. About a third of the country receives an average annual precipitation of less than 260 millimetres/10 inches, whilst only about a fifth receives precipitation of over 760 millimetres/30 inches. In those areas that receive the most rainfall, this also tends to be as a

result of monsoon rains that fall following a prolonged dry season, although Victoria, New South Wales and Tasmania may receive moderate precipitation throughout the year.

Winds tend to be fairly light, although hurricanes and cyclones may affect coastal areas, particularly in the northeast and northwest.

New Zealand

New Zealand possesses a generally temperate climate, although there are regional variations, with warm, subtropical conditions in the far north of the country and alpine climates in the mountains that stretch along both the North and South Islands. These

mountain ranges also have a significant effect on climate and weather throughout New Zealand, acting as a barrier to moist air masses and causing clouds to release that moisture as snow and rain, particularly along the west coast of South Island, which is the wettest region, and also in the Southern Alps, where precipitation, including the water from melted snow, may average over 8000 millimetres/315 inches per year. Elsewhere, most of the country receives an annual average of between about 600 to 1600 millimetres/24 to 64 inches, although the effect of 'rain-shadow' ensures that some areas on the eastern, leeward sides of the mountain ranges may receive less than 406 millimetres/16 inches. With the exception of parts of the south, most of the country experiences increased rainfall in winter, although precipitation tends to be moderate throughout the year, with relatively small seasonal variations.

Similarly, the average annual temperature ranges from about 10–12°C/50–54°F in the south, to around 15–16°C/59–61°F in the north, with little variation between summer and winter overall. However, January and February tend to be the warmest months, and July is generally the coldest. In summer, warm winds may descend from the mountains, sometimes heating inland areas significantly, whilst the coasts tend to benefit from cooling sea breezes.

Pacific Islands

The Pacific Islands are generally grouped into three sub-regions: Melanesia, which lies to the north and north-east of Australia and includes Fiji, Papua New Guinea, the Solomon Islands, Vanuatu, and New Caledonia; Micronesia, which lies further north and east and includes the Caroline Islands, Palau, the Mariana Islands, Guam and the Marshall Islands; and Polynesia, which is comprised of over 1000 islands, roughly occupying a triangular area, with its corners at New Zealand in the west, Hawaii in the north, and Easter Island in the east. Tonga, Samoa, the Cook Islands, and Pitcairn Island are all included within Polynesia.

The climate across this region is essentially humid, and tropical or subtropical. In some areas there may be frequent heavy rains throughout the year, whilst in others there are distinct dry and rainy seasons. Seasonal variations in temperature however are typically slight, tending to average around 27°C/80°F most of the time, and rarely falling lower than 18°C/64°F or exceeding 32°C/90°F, except at higher altitudes. Annual precipitation may be as high as 6400 millimetres/250 inches on some windward island slopes, whilst many isolated, low-lying atolls are highly arid. Hurricanes and Typhoons are frequent across many of the Pacific Islands.

Opposite: A forest with tree ferns in the subtropical area of northern New Zealand.

Below: The Pacific Islands have a tropical or subtropical climate and hurricanes and typhoons are frequent. Hawaii is popular with surfers because of its great waves.

INDEX

BIBLIOGRAPHY

Climate: The force that shapes our world- and the future of life on earth
George Ochoa, Jennifer Hoffman, Tina Tin. Rochdale

Weather: A Visual Guide. Bruce Buckley, Edward J Hopkins, Richard Whitaker, Readers Digest

Weather. Storm Dunlop, Cassell Illustrated

Guide To Weather, Ross Reynolds. Philip's.

How To Identify Weather. Storm Dunlop. Collins

Extreme Weather: A guide and record book. Christopher C. Burt. Norton

Understanding Weather and Climate (4th Edition) by Edward Aguado and James Burt, Prentice Hall

Earth's Climate: Past and Future, William F Ruddiman, Elsevier Science

Meteorology Today: An Introduction to Weather, Climate, and the Environment
(with InfoTrac and Blue Skies CD-ROM) by C. Donald Ahrens Thomson Learning

Weatherwise, P. Eden, Macmillan

How Weather Works, R Chaboud, Thames and Hudson

Fundamentals of weather and climate, R, McIlveen, Stanley Thomas

Weather: The Unltimate Guide to the Element, R Whitaker, HaperCollins

Royal Meterologist Society http://www.royal-met-soc.org.uk

American Meterologist Society http://ametsoc.orG.AMS

Meterological Office, UK http://www.metoffice.com

www.worldclimate.com

World Meteroligcal Organisation http://www.wmo.ch

TORRO (Tornado and Storm Research Organisation http://torro.org.uk

ACKNOWLEDGEMENTS

The Publishers would like to thank the following libraries and photographers
for their kind permission to reproduce their photograph in this book.

Science Photo Library

B & C Alexander 180   Doug Allan 69t ; Jim Amos 224; John Beatty 109t; Ben Nevis Oservatory 223b;George Bernard 97t;   Bettmann 80;   Adrian Bicker 10/11;   Howard Bluestein 209b;   Martin Bond 241,242, 243, 244, 246, 247b ; Peter Bowater 240b;   Brian Brake 53t;   British Antarctic Survey 218, 228, 229t ; Chris Butler 226b;   Cape Grim B.A.P.S 209t;   Jean-Loop Charmet 203t;   CNES 1986 Distribution Spot Image 235t;   Colin Cuthbert 66t ; Michael Donne 197,222;   Bernhard Edmaier 260;   Eye Of Science 85;   Simon Fraser 45t, 67b, 95b, 100, 223t, 225t;   David R.Frazier 103;   Gordon Garradd 46; Geoeye 41t; Geospace 320;   Mark E Gibson 78;   Michael Gilbert 254t, 256t ; Pascal Goergheluck 108 ; Carlos Goldin 262b;   Jim W Grace 85;   David Hay Jones 75t, 208, 210, 215, 255b, 259t ; R.B Hussar 39t; Jet Propulsion Lab. 266 ; John Kaprielian 253 ; Russell Kightley 256br ; Ton Kinsbergen 238;   Paulo Koch 52; Library of Congress 190;   Damien Lovegrove 183b;   Magrath/Folsom 74;   Marine Biological Laboratory 229b; Michael Marten 63b;   Jerry Mason 257b;   Will & Deni Mcintyre 252b;   John Mead 236/237, 240t;   Peter Menzel 162, 209;   Astrid & Hanns-Frieder Michler 143b;   G. Antonio Milani 104 ; Carlos Munos-Yague /Meteo France/Eurelios 220, 221;   Louise Murray 171t;   NASA 33b, 34, 38, 44b, 57b, 71b, 75b, 106, 149t, 159t, 196b, 267b, 270, 273, b275;   NASA/Goddard Space Flight Center 131b; NCAR 86l;   Thomas Nilsen 250, 252t ; NOAA 153, 217t;   Ria Novosti 161, 239t;   NRSC Ltd 44t, 79b; David Nunuk 179b, 183t;   Sam Ogden 40, 211;   David Parker 212, 217b ; Gary Parker 107;   Pekka Parviainen 62, 64, 65, 67t, 68, 71t, 83, 174, 176, 181, 182;   George Post 142 ; Paul Rapson 213;   Jim Reed 77b, 87, 178, 206/207, 216, 272;   Dave Reede 186;   Rev. Ronald Royer 81b;   Chris Sattlberger 130b; Karsten Schneider 93;   Mark A Schneider 66b, 69b, 177;   Peter Scoones 261;   Science Photo Library 26, 41b, 48b, 141, 192, 195, 214, 223, 227, 259b;   Sinclair Stammers 248/249, 256bl;   Kaj R. Svensson 239; Szoenyi Michael 231t, 268;   Sheila Terry 27, 193t;   Mike Theiss / Jim Reed Photography 132t;   Geoff Tompkinson 102;   Mike Umscheid 58;   University Of Dundee 43t, 45b;   Detlev Van Ravenswaay 196t; Jeremy Walker 230;   Art Wolfe 60

Digital Vision

13, 14l, 15, 16, 18b, 19, 20, 22, 24, 25, 29b, 30, 31, 38, 39, 42, 51t, 54, 56, 57t, 58b, 59, 63t, 72, 73, 76, 77t, 79t, 82, 85, 88, 90, 92, 94b, 96, 98, 101b, 110, 115t, 118, 119, 120, 121, 122, 123, 124, 125, 126, 135, 136, 137, 138, 144, 145, 148, 152, 154, 160, 161t, 166, 167, 168, 169, 171b, 172, 176, 180t, 234, 235b, 261, 264, 265, 267, 273t, 278, 282, 283, 284, 285, 286, 287, 288, 290b, 306, 307b, 308, 310b, 315,

Corbis

13b, 14, 17, 18t, 21, 29, 37, 43b, 50, 53b, 78, 80, 91, 101t, 111, 112, 113, 115, 116, 130, 131, 132b, 133, 143t, 149b, 151t, 154t, 155, 156, 157, 158, 165, 170, 184, 187, 190l, 191t, 198, 199, 201b, 203b, 204, 205 (courtesy of the Philadelphia Museum of Art), 226t, 247t, 258, 271t, 280, 281, 289, 290t, 291, 292, 293, 294, 295, 296, 297, 298, 299t, 300, 301, 302b.303t, 304, 305, 307t, 309, 310t, 311, 313, 314

Getty

48t, 49, 70, 81t, 91t, 94t, 99b, 105, 109b, 111b, 114, 128, 140, 146, 150, 188/9b, 189tr, 193b, 197, 199t, 200, 201, 202, 211r, 225b, 231, 243b, 245b, 255t, 262t, 263b, 267t, 274, 276, 285b, 312,